Construction 2018

AA005763

Transportation Research Record: Journal of the Transportation Research Board

Volume 2672, Issue 26

ISBN: 978-1-5108-7830-3

Printed from e-media with permission by:

Curran Associates, Inc.
57 Morehouse Lane
Red Hook, NY 12571

CURRAN ASSOCIATES INC. proceedings.com

Some format issues inherent in the e-media version may also appear in this print version.

Published by Sage Publications USA

ISBN (Print) 978-1-5108-7830-3
ISBN 2018 Printed Set (All Issues) 978-1-5108-7735-1

Printed by Curran Associates, Inc. (2019)

Additional copies of this publication are available from:

Curran Associates, Inc.
57 Morehouse Lane
Red Hook, NY 12571 USA
Phone: 845-758-0400
Fax: 845-758-2633
Email: curran@proceedings.com
Web: www.proceedings.com

Volume 2672 Number 26 December 2018

TRR
JOURNAL OF THE TRANSPORTATION RESEARCH BOARD

Contents

Construction

Articles

EX OFFICIO MEMBERS

---

* Membership as of August 2018.

TRB

TRANSPORTATION RESEARCH BOARD

CORE SPONSORS

Transportation Departments of the 50 States and the District of Columbia

Federal Government

U.S. Department of Transportation
- Federal Aviation Administration (FAA)
- Federal Highway Administration
- Federal Motor Carrier Safety Administration (FMCSA)
- Federal Railroad Administration (FRA)
- Federal Transit Administration
- National Highway Traffic Safety Administration (NHTSA)
- Office of the Assistant Secretary of Transportation for Research and Technology (OST-R)

U.S. Air Force Civil Engineer Center (AFCEC)
U.S. Coast Guard (USCG)
U.S. Department of Energy
U.S. Department of the Interior (DOI)
U.S. Environmental Protection Agency (EPA)

Nongovernmental Organizations

American Public Transportation Association (APTA)
Association of American Railroads (AAR)
California Air Resources Board

The National Academies of
SCIENCES • ENGINEERING • MEDICINE

The **National Academy of Sciences** was established in 1863 by an Act of Congress, signed by President Lincoln, as a private, nongovernmental institution to advise the nation on issues related to science and technology. Members are elected by their peers for outstanding contributions to research. Dr. Marcia McNutt is president.

The **National Academy of Engineering** was established in 1964 under the charter of the National Academy of Sciences to bring the practices of engineering to advising the nation. Members are elected by their peers for extraordinary contributions to engineering. Dr. C. D. Mote, Jr., is president.

The **National Academy of Medicine** (formerly the Institute of Medicine) was established in 1970 under the charter of the National Academy of Sciences to advise the nation on medical and health issues. Members are elected by their peers for distinguished contributions to medicine and health. Dr. Victor J. Dzau is president.

The three Academies work together as the National Academies of Sciences, Engineering, and Medicine to provide independent, objective analysis and advice to the nation and conduct other activities to solve complex problems and inform public policy decisions. The Academies also encourage education and research, recognize outstanding contributions to knowledge, and increase public understanding in matters of science, engineering, and medicine.

Learn more about the National Academies of Sciences, Engineering, and Medicine at **www.national-academies.org.**

The **Transportation Research Board** is one of seven major programs of the National Academies of Sciences, Engineering, and Medicine. The mission of the Transportation Research Board is to increase the benefits that transportation contributes to society by providing leadership in transportation innovation and progress through research and information exchange, conducted within a setting that is objective, interdisciplinary, and multimodal. The Board's varied activities annually engage about 7,000 engineers, scientists, and other transportation researchers and practitioners from the public and private sectors and academia, all of whom contribute their expertise in the public interest. The program is supported by state transportation departments, federal agencies including the component administrations of the U.S. Department of Transportation, and other organizations and individuals interested in the development of transportation.

Learn more about the Transportation Research Board at **www.TRB.org**.

Article

TRR
JOURNAL OF THE TRANSPORTATION RESEARCH BOARD

Transportation Research Record
2018, Vol. 2672(26) 1–10

DOI: 10.1177/0361198118759943
journals.sagepub.com/home/trr
SAGE

Managing Public Communication Strategies in Accelerated Highway Construction Projects

Farzad Minooei[1], Nathaniel Sobin[1], Paul M. Goodrum[1], and Keith R. Molenaar[1]

Abstract

An increase in traveling demands has urged state transportation agencies (STAs) to shorten highway construction project duration, using a variety of accelerated construction techniques and strategies. However, accelerated highway construction projects may impact communities in terms of traffic congestion, safety issues, and other public inconveniences. Effective communication with community stakeholders about these potential inconveniences can provide an opportunity to highlight innovative accomplishments and increase trust between STAs and the traveling public. Therefore, as some studies have shown, effective management of outreach strategies is a key success factor of accelerated construction projects. One of the main goals of this study is to gauge the effectiveness of public outreach tools and strategies in meeting different communication goals, including traffic avoidance, increasing driving awareness, decreasing impatient behavior, and building trust with the public. These tools range from traditional, low-tech methods, such as town hall meetings and static traffic signage, to high-tech methods, such as websites and social media. The data for this research is based on a survey of 71 experts from 40 different STAs. Statistical analyses (analysis of variance) indicate that the effectiveness of outreach tools vary significantly in achieving different communication goals during the construction phase of accelerated highway projects. In addition, the results allowed us to find the most appropriate outreach tools and strategies based on project characteristics and target audience. The findings of this research can help STAs and other public administrators to better manage their outreach activities during the accelerated construction projects.

Steady growth in traffic across the United States during the last decades (*1*) along with an urgent need for transportation infrastructure rehabilitation (*2*) has resulted in an increase in the number of highway construction and maintenance projects. Increasing traffic demand has given rise to congestion, especially in urban areas. Rising traffic congestion, in turn, imposes significant costs on the public in terms of time, money, energy consumption, and environmental impact (*3*). The growing traffic demand along with the current status of roads and bridges indicates that a larger number of road construction projects will be required in the future.

Highway construction itself is a major contributor to the congestion, especially in urban areas. According to the Federal Highway Administration (FHWA), 10% of traffic congestion in the United States is directly related to construction work zones (*4*). As a result, constraints such as time, cost and secondary impacts of highway construction have made schedule acceleration a required element of highway construction projects.

Accelerated highway construction refers to the techniques used to shorten the duration of highway projects. These techniques include both contractual techniques for accelerated construction and innovative construction methods. In addition to their benefit of reducing cost and delivering projects more efficiently, these methods are commonly implemented to decrease project duration and traffic impact (*5*). Project acceleration has become a mantra in many agencies, including the FHWA, where its every day counts (EDC) initiatives have aimed to find and implement innovative solutions for reducing the time to delivery in highway projects, enhancing safety, and protecting the environment (*6*). In parallel, many state Departments of Transportation (DOTs) have also adopted the fundamental principles of the "Get In, Stay In, Get Out, Stay Out" emphasis of the EDC initiatives (*7*).

Empirical evidence provides support for the positive effects of accelerated construction efforts in highway projects. In one study, Lee and Thomas (2007) found that

[1]Department of Civil, Environmental, and Architectural Engineering, University of Colorado Boulder, Boulder, CO

Corresponding Author:
Address correspondence to Farzad Minooei: farzad.minooei@colorado.edu

accelerated construction techniques resulted in a 28% cost reduction and 29% of time-valued savings on a major urban interstate project (*8*). Furthermore, this study demonstrated that appropriate planning and execution of outreach strategies could result in a 20% reduction in construction zone traffic demand and a 50% reduction in maximum peak-hour delay. These findings indicate that well-executed public outreach efforts can decrease the costs in terms of money, time, and resources associated with congestion in accelerated highway construction projects. In addition, this relationship demonstrates the need for understanding which outreach tools and strategies are most effective for different types of projects.

This study employs the experience of public agencies to assess the effectiveness of outreach tools in different scenarios. The results give a comprehensive understanding of the most commonly used outreach strategies by state transportation agencies (STAs) and their effectiveness in addressing different communication goals. These goals consist of reducing traffic demand, increasing driving awareness, decreasing impatient behavior, and building trust with the public. In addition, the results demonstrate the most appropriate outreach tools and strategies based on project characteristics and target audience. The findings of this research can help STAs and other public administrators to better manage their outreach activities during accelerated construction projects.

Background

Communication Theory

Public outreach efforts are based on the principles of communication and mass communication theory. Communication is an iterative process between two or more parties where a message is sent through one or more mediums from a sender to a receiver. The message is encoded and decoded by either sender or receiver with possible noise, which is anything that can interfere with the transmission and understanding of the message (*9*).

Mass communication is similar to communication with the exception that the receiver in mass communication typically consists of a large group of individuals rather than just a single person. Mass communication not only targets multiple individuals, but also provides a level of efficiency. This level of efficiency is measured by considering the use of limited resources such as time and money that are allocated to create the communication (*10*).

The history of mass communication can be broken into two eras. Poster (1995) shows that these two distinct "ages" are delineated by the advent of Internet-based forms of mass communication (*11*). Whereas the first age of mass communication is primarily measured based on the efficiency with which the message could be sent to the greatest number of people, the second age of mass communication is defined by the same measures of efficiency and the degree to which the mass of individuals can reciprocally respond to both the sender and with each other about the original communication (*11–13*).

Public Outreach in Highway Construction Projects

Nearly every STA website provides information related to current traffic and weather conditions on the roads under construction and maintenance by that agency. In addition to websites, transportation agencies also use a range of tools on specific projects. The goals of outreach efforts include four main categories: 1) creating traffic avoidance during construction (e.g. reducing demand); 2) increasing driver awareness in construction zones (e.g. work zone safety); 3) decreasing impatient driver behaviors in construction zones (e.g. speeding and road rage); and 4) building trust with the public (e.g. demonstrating competency, efficiency, or highlighting innovative methods) (*14–17*).

The most focused and recent work on communication efforts in highway construction projects appears in the National Cooperative Highway Research Program (NCHRP) Synthesis 413 (*17*). The synthesis presents 14 categories of communication tools commonly used by transportation agencies for highway construction projects in congested urban areas. These tools range from methods such as traditional town hall meetings to high-tech social media methods such as Twitter. Public notices in the newspaper, project-specific webpages or websites, town hall meetings and flyer distribution have the highest usage rate among outreach tools that are used to involve and inform stakeholder groups.

The NCHRP Synthesis 417 also evaluates the general effectiveness of each tool for communicating with project stakeholders. Interestingly, it shows that the most commonly used forms of engaging and communicating with stakeholders are not always the most effective ones. For example, all survey respondents indicated that public notices in the newspaper are used to engage stakeholders, while this method of communication is ranked the fourth lowest when considering the effectiveness of this form of communication (*17*). One of the few studies that quantitatively gauged the effectiveness of communication efforts during the construction phase of a project is the study of the Interstate 15 project near San Bernardino, California (*18*). The research relates to the effectiveness of automated travel time information, a common form of highway construction communication, for reducing traffic congestion during the construction phase of an accelerated highway construction project. One of the main goals of this study is to provide a convincing quantifiable metric to measure the effectiveness of different public outreach strategies during accelerated construction projects.

Methodology

There are numerous research constraints on measuring the effectiveness of outreach tools. These include a lack of

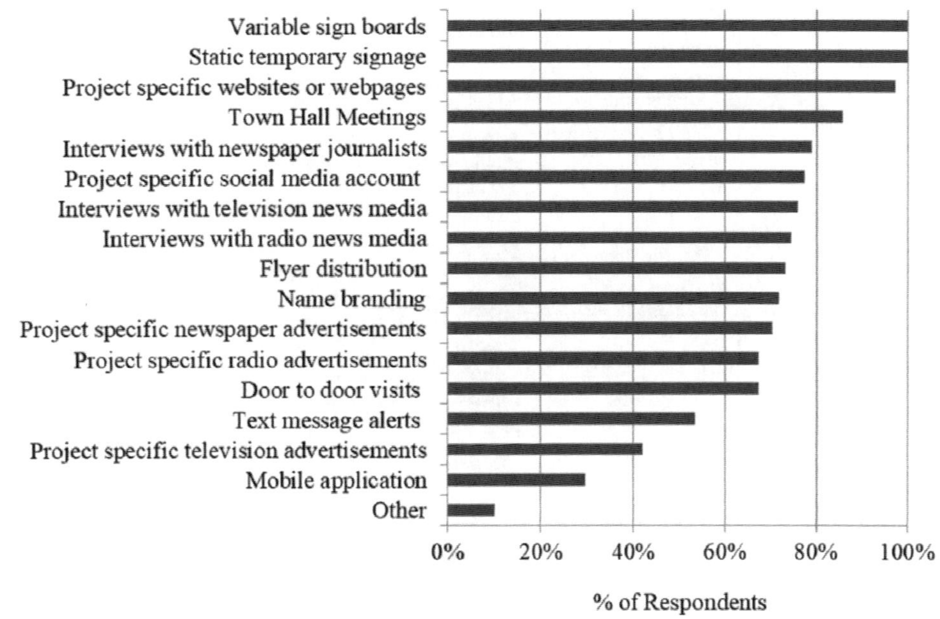

Figure 1. Frequency of usage of outreach tools and strategies.

comparable projects and difficulty in gathering traffic data. Most transportation agencies do not currently collect this type of data for accelerated highway construction projects (*19*). As Warne (2011) states, most agencies gauge the effectiveness of outreach tools through "experience" (*17*). Given these constraints and the data availability, the authors used an online survey methodology to capture the knowledge of experienced practitioners in the area of community outreach practices for accelerated construction projects.

The survey questionnaire design was based on a thorough literature review. An electronic questionnaire was developed and distributed to all 50 STAs in the months of November and December, 2014. After four rounds of data collection, a total of 71 experts from 40 different STAs responded to our survey. We defined expert as individuals actively engaged in the administration and management of a state transportation project. About 71% of the respondents had more than 10 years of work experience in public outreach activities for construction projects. The survey collected information on topics such as different outreach tools and strategies used during accelerated highway construction projects, goals addressed by individual outreach tools, effectiveness of outreach tools to address each goal, selection of outreach tools based on project specifications (traffic impact and capacity loss), and consideration of the types of travelers.

Review of Findings

This section will discuss an overview of outreach tools usage, goals addressed by individual tools, effectiveness of communication strategies, and classification of strategies based on project characteristics and types of travelers.

Outreach Tools and Strategies Used during Accelerated Highway Construction Projects

The literature review revealed 16 different outreach tools and strategies that were commonly used by state transportation agencies. These tools were: static temporary signage; use of variable sign boards; name branding (by using distinctive project name, logo, and catchphrases to enable people to easily recognize any information related to the project); flyer distribution; door-to-door visit by project personnel; town hall meetings; project-specific newspaper, radio, and television advertisements; interviews with newspaper journalists, and radio and television news media; project-specific websites or webpages and social media accounts; text message alerts; and mobile application (*14–17*).

Figure 1 shows the usage of these tools and strategies during the construction phase of accelerated transportation projects among responding agencies. Except for static temporary signage and variable sign boards, which are required by law and used by all STAs, project-specific websites (97%), town hall meetings (86%), interviews with newspapers (79%), social media (77%), and interviews with television (76%) are among the top five outreach tools that are used by STAs. On the other hand, mobile application (54%), project-specific television advertisements (42%), and text message alerts (30%) are less commonly used by STAs.

Outreach Tools and Strategies in Addressing Different Goals

This section presents the tools and strategies that were used most frequently by STAs to achieve each of the four

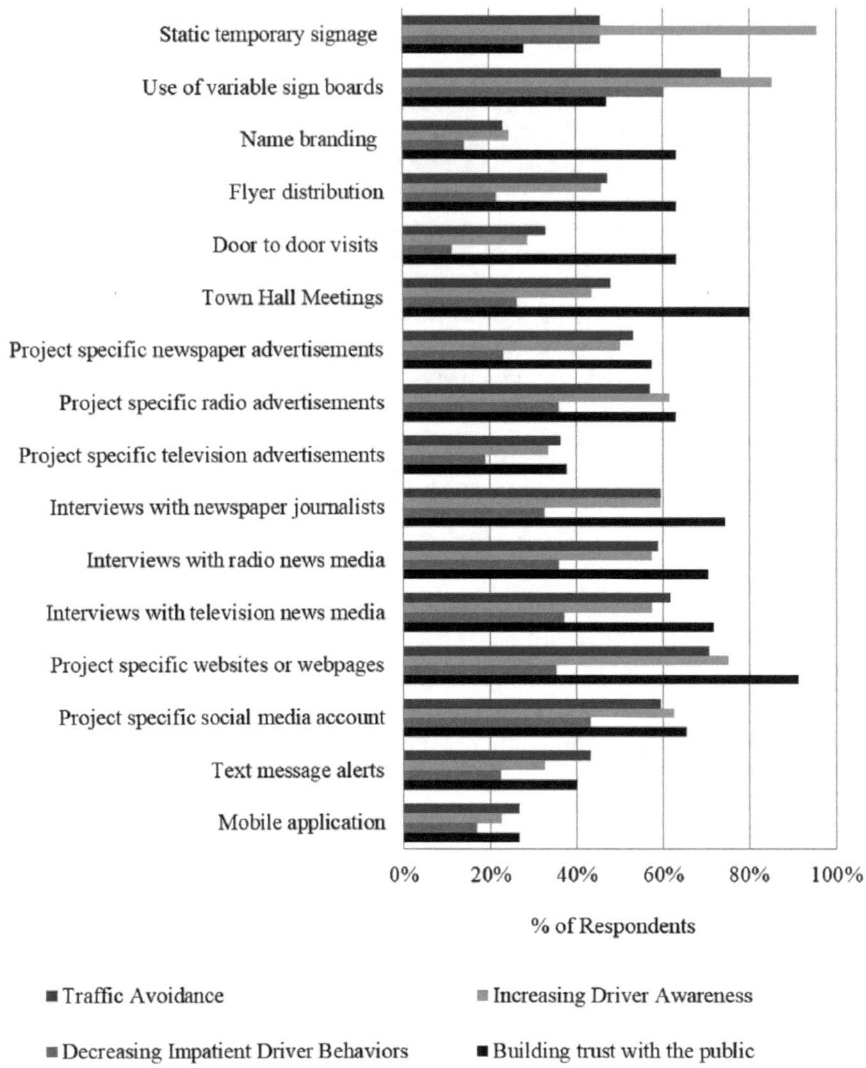

Figure 2. Outreach tools used for addressing different communication goals.

mentioned outreach goals, which include: 1) creating traffic avoidance during construction; 2) increasing driver awareness in construction zones; 3) decreasing impatient driver behaviors in construction zones; and 4) building trust with the public. The results of the chi-square test show that the frequencies of tools usage are statistically significant for each of the four goals of: creating traffic avoidance (χ^2 = 132.84, $p < 0.001$), increasing awareness (χ^2 = 273.81, $p < 0.001$), decreasing impatient behavior (χ^2 = 122.33, $p < 0.001$), and building trust (χ^2 = 207.63, $p < 0.001$). Figure 2 shows the difference in frequency of tool usage in addressing each of the four outreach goals.

Tools and Strategies Used to Create Traffic Avoidance. To create traffic avoidance, except variable sign boards as a communication channel that provide information directly to motorists affected by the project, outreach tools with extensive coverage such as websites, social media, TV, radio and newspapers are used most frequently. Meanwhile, mobile application and name branding are used less frequently.

Tools and Strategies Used to Increase Driver Awareness. As illustrated in Figure 2, on-site communication channels, static signage, and variable sign boards have the highest usage rate among outreach tools that are used to increase awareness. This can be justified by the fact that one of the main functions of these tools is to notify motorists of safety issues in the work zones. The next three most frequently used tools of communication consist of project websites, social media, and radio advertisements. Again, mobile application and name branding are ranked the lowest.

Tools and Strategies Used to Decrease Impatient Driver Behaviors. In order to decrease impatient driver behaviors in construction zones, STAs usually use a combination of on-site outreach tools with immediate effects and tools with wide coverage including social media, interviews with television, and radio advertisements. Relatively low usage rates of most outreach tools for this goal might indicate that STAs do not use a wide range of outreach tools to decrease impatient driver behaviors in construction zones.

Tools and Strategies Used to Build Trust with the Public. Project-specific websites, town hall meetings, interviews with newspapers, television, and radio are the top five communication tools used to attract public trust. Door-to-door visits and name branding are not among the top outreach tools. However, the comparison of their usage frequencies for addressing different goals shows that they are mainly used to build trust with the public. In addition, it is interesting to note that static temporary signage and mobile application are not among the tools that are often used to achieve this goal.

In summary, while there is some consistency in the frequency of use for specific outreach tools across the four strategies, there are differences as well. Variable sign boards, television interviews, social media, and project-specific websites are used relatively frequently across all the four strategies. In contrast, town hall meetings, which are more personal and direct compared to other tools, are used in building trust with the community; their use is relatively rare (less than 50%) across the other three strategies. Although a tool's frequency of usage is an important factor, its effectiveness should also be considered when selecting a tool among different options. This issue will be addressed in the next section.

Effectiveness of Outreach Strategies and Tools

The respondents were asked to rate the effectiveness of outreach tools using a 1–5 Likert scale with a higher rate showing a higher effectiveness to address the related goal. The results of ANOVA show that the effectiveness of different tools is statistically significant for each of the four goals of: creating traffic avoidance ($F = 3.37$, $p < 0.001$), increasing awareness ($F = 5.16$, $p < 0.001$), decreasing impatient behavior ($F = 2.64$, $p < 0.001$), and building trust ($F = 3.00$, $p < 0.001$).

In order to increase the resolution of scores, raw scores were normalized using the min-max method (*20*) based on the equation:

$$P_{norm} = \frac{P_{raw} - P_{rawmin}}{P_{rawmax} - P_{rawmin}} \left(P_{normmax} - P_{normmin}\right) + P_{normmin} \quad (1)$$

where P_{norm} = normalized effectiveness score; P_{raw} = raw effectiveness score; P_{rawmin} and P_{rawmax} = minimum and maximum raw effectiveness values; and $P_{norm\min}$ and $P_{norm\max}$ = minimum and maximum normalized effectiveness values, equal to 1 and 10, respectively. Outreach tools' effectiveness based on normalized scores is demonstrated in Figure 3.

Creating Traffic Avoidance. Interestingly, mobile application and social media are reportedly among the most effective outreach tools in reducing traffic demand. This can be due to the fact that they can provide real-time information to people which allows them to plan in advance of their travels. The next most effective tools for traffic avoidance are project-specific television advertisements, interviews with television and text message alerts, respectively.

Increasing Driver Awareness. Mobile application is also reportedly the most effective tool for increasing driver awareness. In addition, on-site outreach tools, static signage, and variable sign boards are among the tools that have high effectiveness scores. Comparing the results to the frequencies of usage shows that on-site outreach tools are both widespread and effective in increasing driver awareness.

Decreasing Impatient Behavior. The results show that social media, text message alerts, mobile application and variable sign boards have the highest effectiveness in decreasing driver impatient behavior. Although the frequency of usage results shows that STAs use variable sign boards most frequently, these results indicate that there are other effective tools that can be used to address this goal.

Building Trust with the Public. Door-to-door visits and town hall meetings are among the most effective tools to build trust. This can be due to the fact that they allow STAs to have personal contact and direct interaction with people. Social media and mobile applications are also considered as highly effective in building trust among different stakeholders. This can be due to the fact that these tools can increase public engagement via the Internet which in turn broadens public outreach.

To summarize, the results of the survey show that the effectiveness of outreach tools is different across the four communication goals. Interestingly, mobile application is ranked high in addressing the communication goals. This might be due to its real-time and interactive nature. Traditional strategies are still considered effective in addressing some outreach goals. Static signage and variable sign boards are highly effective in increasing driver awareness. In addition, door-to-door visits and town hall meetings are highly effective in building trust with the community.

Project Characteristics and Outreach Strategies

This section provides a framework to suggest outreach tools and strategies based on project characteristics. Two factors of

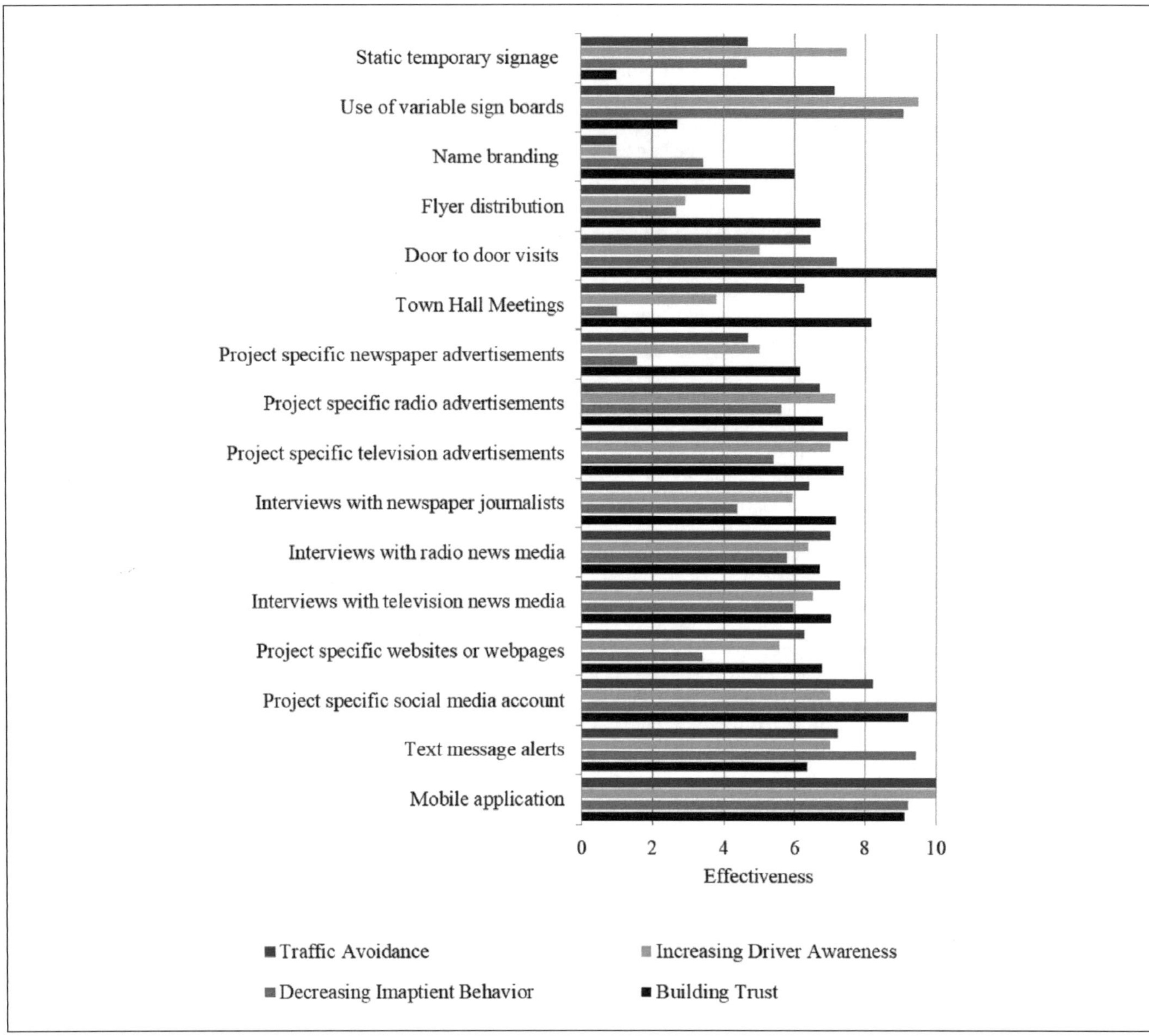

Figure 3. Effectiveness of outreach tools to address different communication goals.

mobility, impact time and levels of closure, are considered in categorizing accelerated highway construction projects. According to the FHWA, the mobility impact time is any period of time that the traffic flow of transportation network is reduced due to on-site construction activities. This metric is classified into five tiers:

- Tier 1 (T1): Traffic impacts within 1–24 hours;
- Tier 2 (T2): Traffic impacts within 3 days;
- Tier 3 (T3): Traffic impacts within 2 weeks;
- Tier 4 (T4): Traffic impacts within 3 months; and
- Tier 5 (T5): Overall project schedule is significantly reduced by months to years (*21*).

The types of closure caused by accelerated highway construction projects can vary greatly as well. In this study, we developed the following levels of highway closure:

- Level 1 (L1): Loss of single shoulder;
- Level 2 (L2): Loss of both shoulders;
- Level 3 (L3): Loss of less than 50% of total travel lanes (typically 1 lane);
- Level 4 (L4): Loss of more than 50% of total travel lanes (typically 2–3 lanes);
- Level 5 (L5): Nighttime closure (single incidence);
- Level 6 (L6): Nighttime closures (multiple incidences);

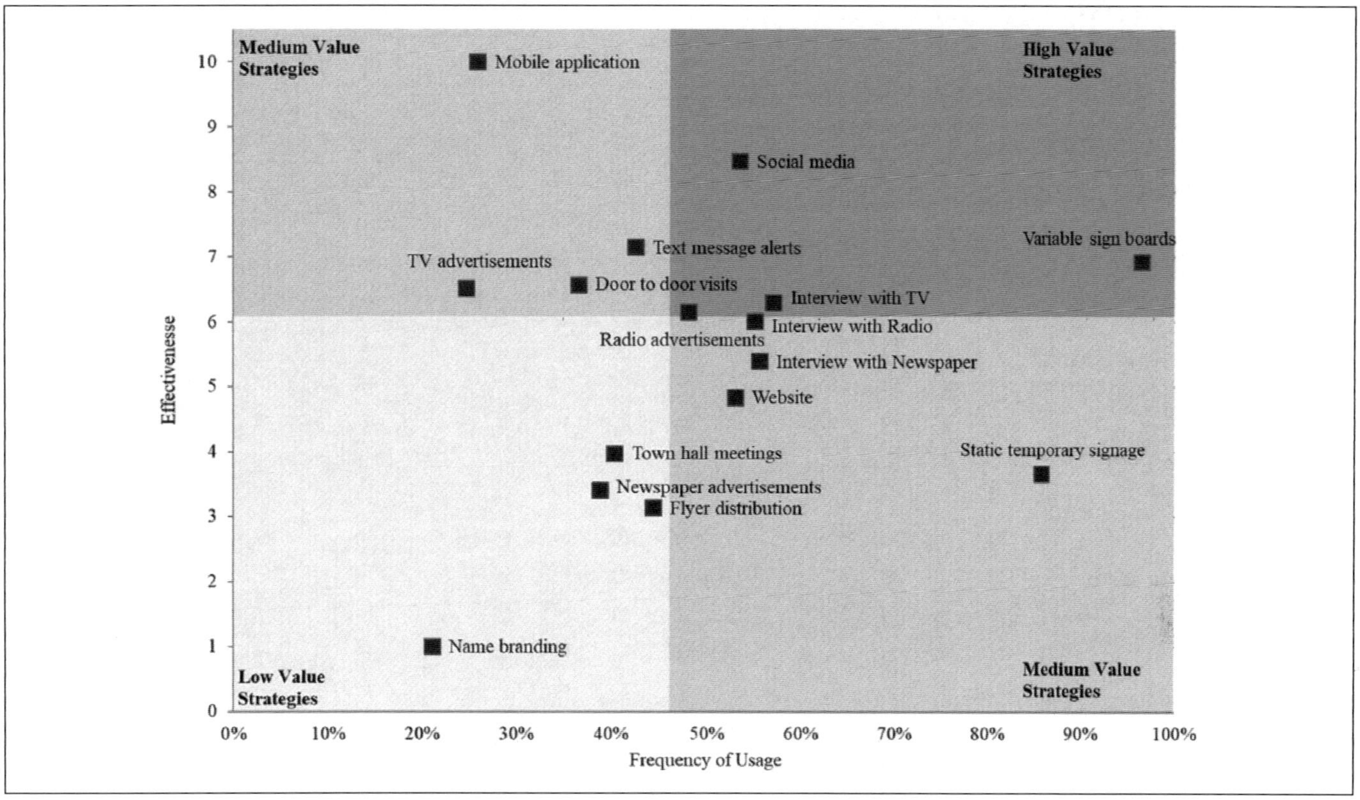

Figure 4. Example of effectiveness-frequency of usage matrix for L7 capacity loss.

- Level 7 (L7): Full closure (less than 1 week's duration);
- Level 8 (L8): Full closure (1 week to 1 month's duration);
- Level 9 (L9): Full closure (1–3 months' duration); and
- Level 10 (L10): Full closure (greater than 3 months' duration).

STAs were asked to select the tools that they use to address projects based on the characteristics mentioned above. The survey results were classified based on both an overall effectiveness score and frequency of usage to provide a coherent framework for recommending outreach tools for different project types. The overall effectiveness score was calculated by taking the average score of effectiveness of each tool to address the four different goals. By using the median score for both effectiveness and frequency of usage, a graphical plot was developed for each project characteristic and divided into four quadrants to identify three-tier strategies. For example, Figure 4 illustrates the plot for a project with L7 capacity loss. Three-tier strategies are defined for each project type as follows: High Value: strategies with high effectiveness and high frequency; Medium Value: strategies with either high effectiveness and low frequency or low effectiveness and high frequency; and Low Value: strategies with low effectiveness and low frequency. The summary of the results for different project characteristics are shown in Table 1. Note that since static temporary signage is mandatory, it was always categorized as a High Value strategy.

Target Audience and Outreach Strategies

In addition to addressing project characteristics in relation to project impact and capacity loss, the study also found that 87% of the responding agencies considered types of travelers (commuter (C), non-commuter (NC), and commercial trucks (CT)) when they thought about outreach tools they wanted to use. Commuters are more likely to be informed about travel but they are less flexible about different aspects of their trip. By contrast, non-commuters are usually harder to access but more likely to react to the messages about changing the timing of a trip and their destination. Finally, commercial drivers have different needs regarding work zone information because of their tight schedule, transportation of oversize or hazardous materials, and overnight travel (*14*).

The matrix approach has been used to classify outreach tools and strategies relevant to different types of travelers. Table 1 summarized the results for all types.

Discussion of Findings

By comparing the results of frequency of usage with effectiveness of outreach tools, it becomes apparent that the most commonly used methods employed to reach specific goals

Table 1. Classification of Outreach Strategies Based on Project Characteristics and Types of Travelers

	Project types													Target audience	
	Project impact					Capacity loss									
Tools & strategies	T1	T2	T3	T4	T5	L1	L2	L3	L4	L5	L6	L7 & L8	L9 & L10	C & NC	CT
Static temporary signage	H	H	H	H	H	H	H	H	H	H	H	H	H	H	H
Variable sign boards	H	H	H	H	H	H	H	H	H	H	H	H	H	H	H
Name branding	L	L	L	L	M	L	L	L	L	L	L	L	L	L	L
Flyer distribution	L	L	L	M	M	M	M	L	L	L	M	L	M	L	L
Door to door visits	M	M	M	H	M	M	M	M	M	M	M	M	M	M	M
Town hall meetings	L	L	L	M	M	L	L	L	L	L	L	L	M	M	L
Newspaper advertisements	L	L	L	L	L	M	L	L	L	L	L	L	L	L	L
Radio advertisements	H	H	H	M	M	H	H	H	M	H	H	H	M	H	H
TV advertisements	M	M	M	M	M	M	M	M	M	M	M	M	M	M	M
Interviews with newspaper	L	M	M	M	M	L	L	L	M	M	M	M	M	M	M
Interviews with radio	M	M	M	M	M	L	L	M	M	M	L	M	M	L	M
Interviews with TV	H	H	H	H	H	L	M	H	H	M	M	H	H	H	H
Website	L	L	L	L	M	M	M	M	M	M	M	M	M	M	M
Social media	H	H	H	H	H	H	H	H	H	H	H	H	H	H	H
Text message alerts	H	H	H	M	M	H	H	H	H	H	H	M	M	M	H
Mobile application	H	M	M	M	M	H	H	M	M	M	M	M	M	M	M

Note: H = highly recommended (high value strategies); M = moderately recommended (medium value strategies); L = recommended (low value strategies); T1 = traffic impacts within 1–24 hours; T2 = traffic impacts within 3 days; T3 = traffic impacts within 2 weeks; T4 = traffic impacts within 3 months; T5 = overall project schedule is significantly reduced by months to years; L1 = loss of single shoulder; L2 = loss of both shoulders; L3 = loss of less than 50% of total travel lanes (typically 1 lane); L4 = loss of more than 50% of total travel lanes (typically 2–3 lanes); L5 = nighttime closure (single incidence); L6 = nighttime closures (multiple incidences); L7 = full closure (less than one week duration); L8 = full closure (1 week to 1 month's duration); L9 = full closure (1 to 3 month's duration); L10 = full closure (greater than 3 month's duration); C = commuter; NC = non-commuter; CT = commercial trucks.

Table 2. Three-Tier Strategies for a Project with T3 Traffic Impact, L6 Capacity Loss, and Primary Commuter Traffic

Outreach Tools and Strategies
High value
Interviews with TV
Radio advertisements
Social media
Static temporary signage
Text message alerts
Variable sign boards
Medium value
Door to door visits
Flyer distribution
Interviews with newspaper
Interviews with radio
Mobile application
Town hall meetings
TV advertisements
Website
Low value
Name branding
Newspaper advertisements

are not always the most effective ones. One interesting discrepancy is mobile applications. Although, mobile applications are not used by many respondents, they are categorized among highly effective tools by those respondents who used them to communicate with stakeholders and the public. Also, text messaging was ranked among the top five effective tools in creating traffic avoidance, increasing driver awareness, and decreasing impatient behavior. In 2013, more than 90% of adults in the United States owned a cell phone and 56% of the cell phones were smart phones (*22*). These facts along with the results of our study provide evidence for the effectiveness of cell phones as communication channels in accelerated construction projects. However, this recommendation should be considered with caution due to the concerns regarding the use of cell phones while driving.

Another point is that social media is frequently used by STAs to achieve most communication goals and is also reported as highly effective in achieving the related goals. This finding is supported by the survey conducted by the American Association of State Highway Transportation Officials (AASHTO) Subcommittee on Transportation Communication which indicates that the use of social media for communication is constantly on the rise by state DOTs (*22*).

Table 1 can be used to recommend outreach tools and strategies during the construction phase of an accelerated highway project with multiple characteristics. As an example, it may be of interest to know the recommended strategies for a project with T3 traffic impact, L6 capacity loss, and primary commuter traffic. Table 2 lists the selected parts of the table that are related to this specific project and unites different sets of strategies in three tiers. The authors used a conservative approach in placing tools, which means that in cases where one tool belonged to two different tiers, it was placed in the higher tier. In this example, town hall meetings are in both Low Value and Medium Value tiers, but were classified in Medium Value strategies. It should be noted that High Value strategies are highly effective and are most frequently used based on the survey data. They are highly recommended to be applied for the accelerated projects. Medium Value strategies are either highly effective or most common but using them in different projects needs a more thorough analysis. Factors such as outreach goals, budget, internal and external resources, and work zone location may influence the decision to employ them.

Conclusions

Accelerated highway projects may have both negative and positive effects on society during the construction phase. Effective communication with stakeholders can not only reduce negative side effects of the project such as traffic congestion and accidents in the work zone, but also provide opportunities for building trust between STAs and the public by showing competence and efficiency.

This research examines the recent experiences of STAs in using different outreach tools and strategies during the construction phase of accelerated transportation projects. Findings from the survey of 71 experts indicate that with exception to static temporary signage and variable sign boards which are used in almost every project, project-specific websites, town hall meetings, interviews with newspapers, social media, and interviews with television are used most frequently by STAs during the construction phase of accelerated projects. Moreover, the frequency of usage of outreach tools to address four main communication goals is analyzed.

The effectiveness of communication tools and strategies are also gauged in this study. Statistical analysis of the survey reveals that outreach tools and strategies vary significantly in achieving each communication goal. The analysis of data shows that social media is ranked high in both effectiveness and frequency of usage, which justifies the increase in the use of social media by STAs. In addition, the survey results indicate that the most commonly used methods employed to reach specific goals are not necessarily the most effective ones. One interesting case is mobile application which currently is among the less frequent outreach tools but was reported as highly effective to address most communication goals.

The results of the survey also enable us to provide a framework to recommend a range of outreach strategies are categorized based on project characteristics and types of traveler: communication strategies based on traffic impact of the project, capacity loss imposed by construction operations, and types of travelers targeted for communication. Three tiers of strategies are recommended in each situation. This classification also allows the recommendation of three-tiered strategies for projects with multiple characteristics. However, other factors such as outreach goals, budget, internal and external resources, and work zone location should be considered by decision-makers when employing these strategies.

Acknowledgments

This research was funded by Transportation Construction Management (TCM) pooled-fund program and its participating members. The opinions expressed herein are those of the authors and not necessarily of the TCM members.

References

1. Federal Highway Administration (FHWA). *Travel Volume Trends*. FHWA, Office of Highway Policy Information, May 2015 Report. http://www.fhwa.dot.gov/policyinformation/travel_monitoring/15maytvt/15maytvt.pdf. Accessed July 20, 2017.
2. American Society of Civil Engineers (ASCE). *2013 Report Card for America's Infrastructure*. ASCE. http://www.infrastructurereportcard.org/a/#p/grade-sheet/gpa. Accessed July 20, 2017.
3. Schrank, D., B., Eisele, and Lomax, T. *TTI's 2012 Urban Mobility Report*. Texas A&M Transportation Institute, December 2012. http://d2dtl5nnlpfr0r.cloudfront.net/tti.tamu.edu/documents/mobility-report-2012.pdf. Accessed July 20, 2017.
4. Federal Highway Administration (FHWA). *Operations Story*. FHWA. http://www.ops.fhwa.dot.gov/aboutus/opstory.htm. Accessed July 20, 2017.
5. American Association of State and Highway Transportation Officials (AASHTO). *AASHTO Guide for Design-Build Procurement*. AASHTO, Washington, D.C., January 2008.
6. Federal Highway Administration (FHWA). *Every Day Counts*. http://www.fhwa.dot.gov/everydaycounts/. Accessed July 20, 2017.
7. Mistry, V., and Mangus, L. Get In, Stay In, Get Out, Stay Out. *Public Roads*, FHWA-HRT-2006-001, Vol. 70, No. 3, 2006. http://www.fhwa.dot.gov/publications/publicroads/06nov/01.cfm. Accessed July 20, 2017.
8. Lee, E. B., and D. K. Thomas. State-of-Practice Technologies on Accelerated Urban Highway Rehabilitation: I-15 California Experience. *Journal of Construction Engineering and Management*, Vol. 133, No. 2, 2007, pp. 105–113. https://doi.org/10.1061/(ASCE)0733-9364(2007)133:2(105).
9. Project Management Institute (PMI). *A Guide to the Project Management Body of Knowledge*, 4th ed. Project Management Institute, Newtown Square, Pa., 2008.

10. Danesi, M. *Encyclopedia of Media and Communication*. University of Toronto Press, Toronto, Canada, 2013.
11. Poster, M. *The Second Media Age*. Blackwell Publishers, Inc., Cambridge, MA, 1995.
12. Holmes, D. *Communication Theory: Media, Technology, and Society*. SAGE, Thousand Oaks, CA, 2005.
13. McQuail, D. *McQuail's Mass Communication Theory*, 6th ed. SAGE, Thousand Oaks, Ca, 2010.
14. Mallett W. J., J. Torrence, and J. Seplow. *Work Zone Public Information and Outreach Strategies*. Federal Highway Administration (FHWA), FHWA-HOP-05-067, November 2005. http://www.ops.fhwa.dot.gov/wz/info_and_outreach/public_outreach_guide.pdf. Accessed July 20, 2017.
15. Federal Highway Administration (FHWA). *Work Zone Operations – Best Practices Guidebook*. FHWA, FWHA-HOP-07-131, October 2007.
16. Maryland State Highway Administration. *Public Information and Outreach Plans*. October 2006. http://www.roads.maryland.gov/OOTS/11PIOforTMPRev1.pdf. Accessed July 20, 2017.
17. Warne, T. *NCHRP Synthesis 413: Techniques for Effective Highway Construction Projects in Congested Urban Areas*. HRB, National Research Council, Washington, D.C., 2011.
18. Lee, E. B., C. Kim, and J. T. Harvey. Implementation of Automated Travel Time Information and Public Reaction on Urban Highway Rehabilitation. *Journal of Transportation Engineering*, Vol. 132, No. 10, 2006, pp. 808–816. https://doi.org/10.1061/(ASCE)0733-947X(2006)132:10(808).
19. Hallmark, S., J. Turner, and C. Albrecht. *Synthesis of Work Zone Performance Measures*. Iowa State University, Institute for Transportation (InTrans). Project 12-436, TPF 5(081), September 2013. http://www.intrans.iastate.edu/research/documents/research-reports/work-zone_perf_msrs_w_cvr2.pdf. Accessed July 20, 2017.
20. Han, J., M. Kamber, and J. Pei. *Data Mining: Concepts and Techniques*. Morgan Kaufmann, San Francisco, 2006.
21. Culmo, M. P. *Accelerated Bridge Construction – Experience in Design, Fabrication and Erection of Prefabricated Bridge Elements and Systems*. FHWA, FHWA-HIF-12-013, November 2011.
22. American Association of State and Highway Transportation Officials (AASHTO). *State DOT Social Media Survey*. AASHTO Subcommittee on Transportation Communications, September 2013. http://communications.transportation.org/Documents/Social%20Media%20Survey_final_Oct2013.pdf. Accessed July 20, 2017.

The Standing Committee on Construction (AFH00) peer-reviewed this paper (18-03581).

Article

TRR
JOURNAL OF THE TRANSPORTATION RESEARCH BOARD

Transportation Research Record
2018, Vol. 2672(26) 11–20

DOI: 10.1177/0361198118758052
journals.sagepub.com/home/trr
SAGE

Unit Price or Lump Sum? A Stochastic Cost-Based Decision-Making Tool for Design-Bid-Build Projects

Mohamed Khalafalla[1] and Jorge Rueda-Benavides[1]

Abstract

The appropriate selection of procurement tools and strategies is a key factor in the successful completion of construction projects. Despite the increasing use of alternative project delivery methods by the public agencies, the traditional design–bid–build (DBB) approach is still the most used and accepted project delivery method in the US. The purpose of this study is to present an innovative tool that introduces an objective criterion into the selection of resurfacing projects for design–bid–build —lump sum (DBB-LS) projects. This criterion consists of a comparison of the stochastic construction cost estimate for a candidate project if procured under DBB-LS versus its stochastic cost estimate if a unit price (UP) approach is used. The proposed decision-making tool was developed using non-linear regression techniques, Monte Carlo Simulation, and data from 86 resurfacing projects completed by the Florida Department of Transportation (FDOT) between January 2015 and March 2017: 63 UP and 23 LS projects. To facilitate the use of the tool by decision makers, the mathematical functions involved in the cost comparison between these two approaches have been comprised into a nomogram, allowing a quick approximation of the probability of having a lower cost under each compensation approach and the potential savings or extra costs of using an LS provision. The nomogram also allows the estimation of potential LS cost implications under different confidence levels, providing FDOT with the ability to make decisions at different levels of risk.

Public owners in the transportation construction industry have been increasingly using alternative project delivery methods (e.g. design-build; construction manager at risk; or partnership) to maintain, repair, and expand the transportation infrastructure in the U.S. The implementation of these methods is part of the efforts made by federal and state transportation agencies (STAs) to cope with the tight deadlines and even tighter budgets of today's construction industry. Alternative project delivery methods have proved their effectiveness in improving project performance with reference to quality, cost, and project duration (*1*).

However, STAs recognize that alternative procurement strategies are not always the most suitable approach. In fact, most transportation construction and maintenance projects are currently being procured through traditional design–bid–build (DBB) contracting (*2*). For instance, between January 2015 and March 2017, the Florida Department of Transportation (FDOT)—one of the leading STAs in the use of alternative contracting methods—awarded more than 71% of all its projects using a traditional DBB approach (*3*). Therefore, any efforts toward the improvement and optimization of DDB practices are expected to have a considerable impact on STAs' construction and maintenance programs and to contribute to the appropriate use of taxpayers' money.

In DBB contracting, design is fully accomplished by the STA, using either in-house or consultant designers, before advertising and awarding a separate construction contract (*1*). As occurs with any other project delivery method, DBB contracts can be tailored, to some extent, to match specific project needs. It is done through the combination of procurement procedures intended to address different administrative and management aspects of the project (*4*). In the case of procedures to compensate general contractors, a construction contract on a DBB project could provide for compensation based on the actual amount of work performed [unit price (UP)], on a lump sum (LS) proposed by the contractor, or on the actual costs incurred by the contractor plus a fee to allow for a profit (cost reimbursable or cost-plus). It would depend on the specifics of the project or the preferences of the agency or both. Contractual obligations and exposure to risk vary depending on the selected compensation provision, meaning that contractors must develop a different pricing strategy for each compensation approach. It also means that when selecting a

[1]Auburn University, Auburn, AL

Corresponding Author:
Address correspondence to Mohamed Khalafalla: mzk0094@auburn.edu

compensation approach, STAs should consider the cost implications associated with each alternative. Nevertheless, there is a lack of mechanisms to factor construction cost estimates into the selection of procurement strategies—a widespread problem across all project delivery methods and contracting approaches (*1, 4, 5*). "[B]ecause no adequate and systematic method exists to evaluate how project delivery methods and contracting approaches have affected costs, it is difficult to validate the financial impacts of their use" (*4*).

This paper describes the development of a cost-based decision-making tool intended to assist FDOT with the selection of contractor compensation provisions in DBB projects. More specifically, the proposed tool facilitates the comparison of stochastic cost estimates associated with the use of the two main compensation approaches used by FDOT in DBB construction and maintenance projects: UP and LS. The use of stochastic estimates allows accounting for cost escalation trends associated with each of the two approaches. The tool was developed using non-linear regression techniques, Monte Carlo Simulation, and data from 86 resurfacing projects completed by FDOT between January 2015 and March 2017: 63 UP and 23 LS projects.

The paper provides enough information for the development of spreadsheets to facilitate the use of the tool by decision makers. Additionally, the mathematical functions involved in the cost comparison between UP and LS have been comprised into a nomogram to facilitate the use of the tool by different types of decision makers and in different environments. The nomogram allows a quick approximation of the probability of having a lower cost under each compensation approach and the potential savings or extra costs of using an LS provision. The nomogram also facilitates the assessment of LS cost implications under different confidence levels, allowing FDOT to make decisions under different levels of risk. Even though the proposed tool is only applicable to resurfacing projects awarded by FDOT, the methodology presented in this paper could be replicated for other agencies and types of projects as well as include other compensation approaches.

Background

Under FDOT's current contracting program, LS contracting is classified as an alternative approach to DBB (*5*). However, for the purposes of this paper, LS is considered as a contract provision that can be used in DBB contracts as a substitute to the usual UP bidding/compensation approach. This assumption is made because LS contracts awarded by FDOT do not alter the traditional DBB contractual relationships between owners, designers, and contractors (*6*). The design and construction phases are still performed in sequence and by different entities. LS provisions are also used by FDOT in design-build contracts; however, this study is only focused on DBB-LS contracts, which are usually labeled by FDOT as "lump sum contracts" (*7*).

Besides dictating the procedures to compensate contractors for the work performed, payment provisions also define the bidding approach to be used in construction and maintenance contracts. In UP contracts, bidders are required to submit UPs for all pay items and bid quantities listed in the solicitation documents, and the contract is usually awarded on a low-bid basis. The selected contractor is then compensated through partial payments based on actual quantities of work measured by the owner (or its representative) and the UPs submitted by the contractor (*8*). On the other hand, in LS contracts, contractors prepare and submit an LS bid based on plans and specifications advertised by the owner in the solicitation documents (*9*). An LS bid is a fixed amount of money in exchange of which the contractor agrees to perform all work described in the contract documents. This amount includes all material, labor, equipment, overhead, and profit (*5, 10*). As in UP contracts, DBB-LS contracts are commonly awarded to the lowest bidder, but in this case, the selected contractor is compensated based on the LS bid and a payout schedule agreed on by the owner and the contractor (*7, 8, 11*). FDOT has made clear that "Lump Sum contracts are not fixed price" (*7*). It means that LS bids submitted by contractors are susceptible to changes because of changes in the scope of work, extra work, or unforeseen conditions. FDOT's LS contract documents include predetermined UPs for key work items to facilitate the pricing of change orders and extra work (*7*).

Most authors and STAs agree that one of the main benefits that motivates the use of LS contracts is the reduction of contract administration costs and efforts related to quantity verification and measurement (*5–8, 11, 12*). It allows field inspectors to pay more attention to aspects that ensure the final product will meet minimum quality and performance standards (*8, 11*).

LS Advantages (6–11)

- Reduce contract administration costs and efforts related to quantity verification and measurement.
- Simplify the payment process.
- Allow inspectors to focus on quality and performance standards.
- Incentivize contractors to implement better cost control measures.
- Greater flexibility to contractors to select construction means and methods.

LS Disadvantages (6–11)

- Greater difficulty in identification, pricing, and negotiation of change orders and extra work.
- Proposals may include higher contingencies.
- Require higher design quality to avoid change orders and potential disputes/claims.

Benefits offered by LS contracts are the result of moving procurement practices toward a more "all-inclusive" approach; however, this also seems to be the cause of their main disadvantage: the greater difficulty in managing and negotiating extra work and change orders. The absence of UPs for specific materials or activities makes it difficult for both STAs and contractors to price change orders and extra work as well as to identify legitimate deviations from the original scope of work contained in the LS bid (*7, 8*). This situation may force contractors to increase contingencies in their price proposals to compensate for the greater risk. To mitigate the impact and likelihood of the occurrence of the identified disadvantages, DBB-LS users have developed their own project selection criteria. Examples of these criteria include projects with:

- A well-defined scope of work for all parties (*9, 10*);
- Low risk of unforeseen conditions (*9, 10*);
- Low possibility for scope changes during all project phases (*9, 10*);
- Low possibility for contractors to provide less than the required quantities of work (*10*).

As an additional strategy to facilitate the decision-making process on whether to use LS provisions instead of UP approach, STAs have included in their LS guidelines examples of specific types of projects that may be suitable for LS.

Most Suitable Types of Projects for LS Contracting (9–11)

- Bridge painting.
- Bridge projects (with limited earthwork or pile driving).
- Fencing/guardrail installation.
- Intersection improvements.
- Landscaping.
- Lighting.
- Simple milling and resurfacing.
- Minor road widening (with limited earthwork).
- Sidewalks.
- Signing.
- Signalization.
- Simple transportation enhancement.

Least Suitable Types of Projects for LS Contracting (9–11)

- Urban construction/reconstruction.
- Rehabilitation of movable bridges.
- Projects with subsoil earthwork.
- Concrete pavement rehabilitation.
- Major bridge rehabilitation/repair.

Research Approach

The research methodology followed in this study is illustrated in Figure 1. The first step in the study was a comprehensive literature review to gain an insight into STAs' DBB contracting practices and contractor compensation procedures as well as to identify key data and information to be required throughout the study. All identified relevant data and information was then gathered by the research team, including historical bid data from all projects awarded by FDOT between January 2015 and March 2017. Bid data was collected from bid tabulations published by FDOT (*12*).

The primary objective of this study is to develop a methodology to identify suitable projects for LS contracting for all types of work procured by FDOT. The application of the methodology presented in this paper is for a single type of work. However, the same process can be replicated for projects with different scopes of work. The methodology must be applied separately to each type of work and with different data sets. This is a data-driven approach; therefore, the effectiveness of its application depends on the quality and amount of available data. Thus, the next step in the study was a review of the available data to identify a suitable type of work to effectively demonstrate the application of the proposed decision-making tool. The authors found that available data from resurfacing projects was the most suitable for this project with reference to data quality and quantity. Resurfacing is defined as a work to add asphalt layers to an existing pavement to extend the service life of the road (*13*).

Data from 338 resurfacing projects were extracted from the original data set. The data were carefully inspected and further reduced based on the needs of the study. The following criteria were used for the final selection of projects:

- Only projects with scopes of work exclusively focused on resurfacing activities (projects including other types of work besides resurfacing were discarded).
- Only completed projects (projects in which final payments have been made to contractors so as to have access to actual construction costs).
- Only projects with uniform road cross section along the project to facilitate the calculation of construction costs per lane mile.

Google Earth was used to review the road cross section of all 338 projects and to verify the number of lane miles per project. A total of 86 resurfacing projects (63 UP and 23 LS projects) met the criteria listed above and were used toward the development of the LS project selection model. It should be noted that the criteria to select suitable projects for this study was used only for modeling purposes. The proposed tool is applicable to all resurfacing projects for which the total number of lane miles has been determined.

The data analysis process for the selected 86 resurfacing projects was divided into four parts (see Figure 1). In the first

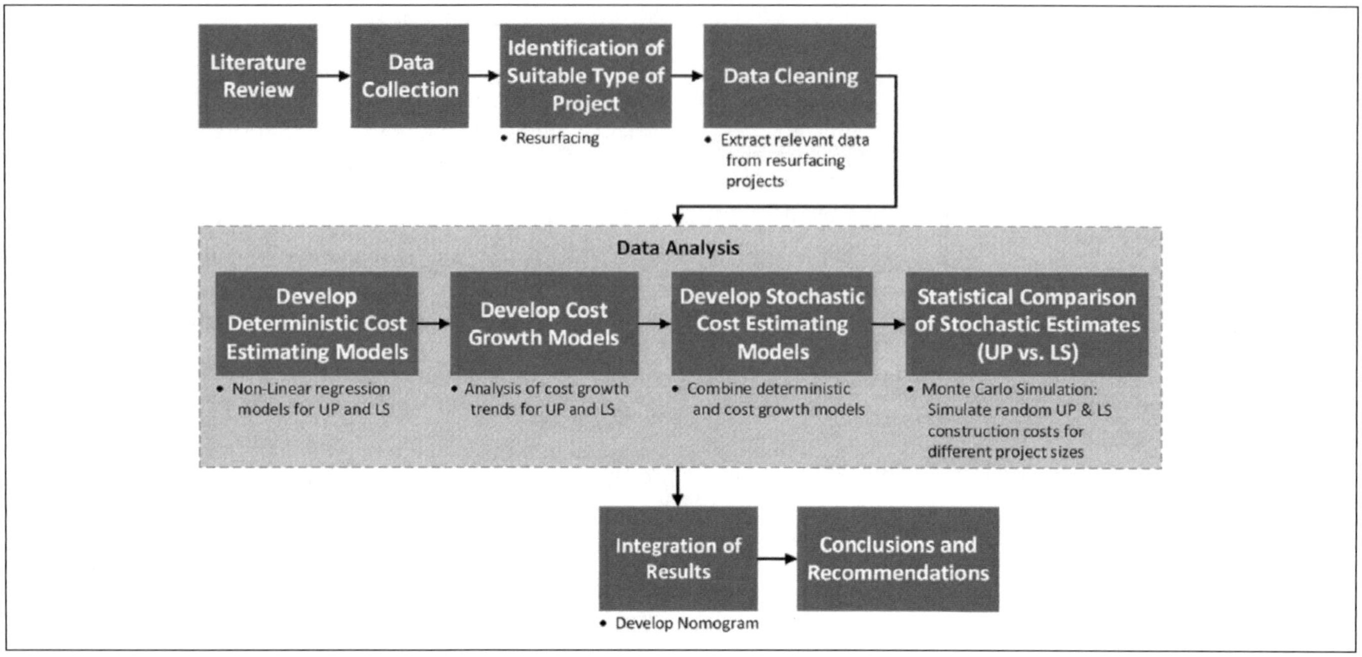

Figure 1. Research methodology.

part, the authors used non-linear regression techniques to develop deterministic cost estimating models based on the total number of lane miles for both UP and LS contracts. The second part of the data analysis process consisted of modeling the expected cost growth for each compensation approach based on the quantitative analysis of cost estimating uncertainty. Deterministic cost estimating models and cost growth models were then integrated to produce two stochastic cost estimating models: one for UP and one for LS projects. A stochastic approach was used in this study to account for the unavoidable cost estimating uncertainty faced by estimators and project managers in the construction industry.

The stochastic cost estimating models were then used to determine the probability of UP costs exceeding LS costs for different project sizes (size measured in lane miles) using Monte Carlo simulation techniques. Finally, simulation results were used to develop a nomogram that comprises the entire methodology into a two-dimensional chart. The development and use of the nomogram will be explain in more detail throughout the paper.

Data Analysis

This section presents the results for each of the four parts of the analysis process illustrated in Figure 1.

Deterministic Construction Cost Estimates. The deterministic cost estimating models for UP and LS were developed using historical bid data and non-linear regression techniques to correlate actual total bid prices per lane mile submitted by successful contractors with the total number of lane miles to be resurfaced. The total number of miles in resurfacing projects are usually known early during the planning phase. Therefore, these deterministic models provide early estimates for UP and LS projects based on the total number of lane miles. The use of early construction cost estimates is a key factor in the proposed methodology as "the decision to use the Lump Sum Contracting Technique on a project should be made during the scope development process, rather than during or after the design process" (*9*).

Figures 2 and 3 illustrate the deterministic construction cost estimating models for UP and LS contracts, respectively. The regression equation that best fits each data set is shown in the top-right corner of each figure. Both curves show an inverse relationship between total lane miles and cost per lane mile, which was expected by the authors. Larger amounts of work in construction projects are usually paid at lower rates per unit of work (*14*). The deterministic model for UP project (Figure 2) shows a stronger correlation between total bid price per lane mile and total number of lane miles than that in the LS model. This could be because the UP model was developed with a considerably higher number of data points. Sixty-three UP projects are shown in Figure 2, whereas the model for LS project was built with 23 projects. This study assumes the non-linear regression model shown in Figure 3 is a good approximation of the model that would be obtained with a larger sample of projects.

An interesting observation was made when plotting both the UP and the LS deterministic cost estimating models in the same chart, as shown in Figure 4. This figure shows that, on average, resurfacing projects for less than 21 lane miles are expected to have a lower cost per mile if LS provisions are used. The

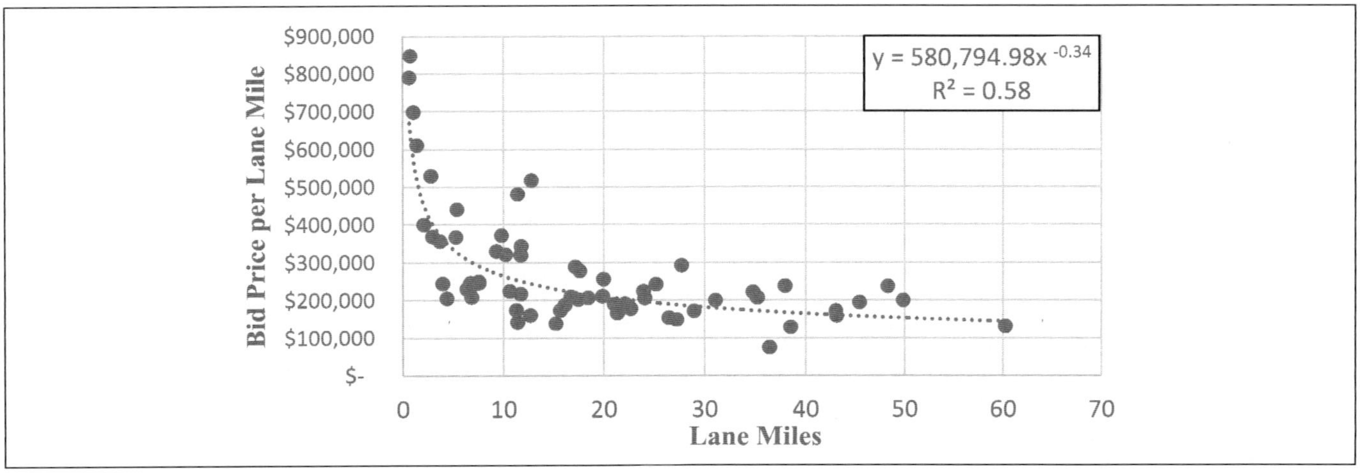

Figure 2. UP deterministic cost estimating model.

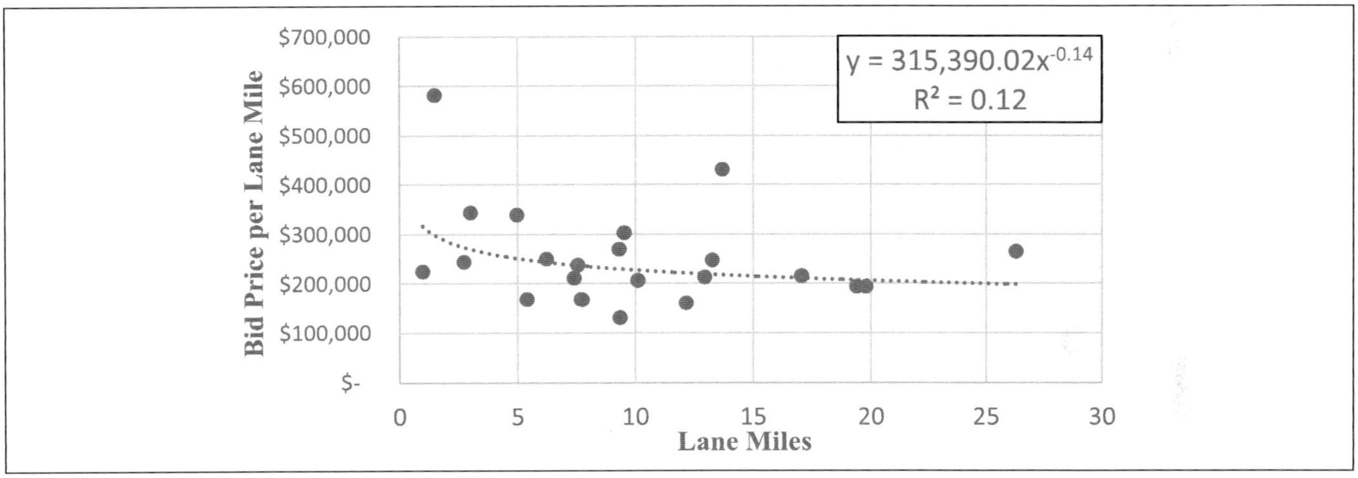

Figure 3. LS deterministic cost estimating model.

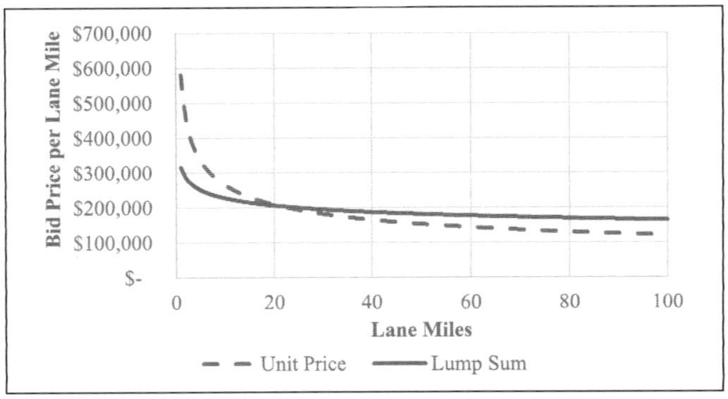

Figure 4. UP and LS deterministic cost estimating models.

comparison of this break point against the data from LS projects used in this study revealed that only one of the 23 LS resurfacing projects was for more than 21 lane miles. It could be an indicator of the potential cost-effectiveness of FDOT's current subjective LS project selection procedures, which are based on selecting "simple" projects for LS according to three criteria: well-defined scope, low risk of unforeseen conditions; and low possibility of scope change (*9*). This is an important observation; however, it should be noted that these models were built with total bid prices and do not reflect actual costs paid by FDOT. Likewise, these are deterministic models, which are average representations of the cost behavior of UP and LS projects. It does not mean that all resurfacing projects for less than 21 lane miles will have lower costs for FDOT. Deterministic models ignore the cost risk inherent in construction cost estimating and do not provide the decision maker with the ability to consider multiple case scenarios. The proposed methodology is intended to be an improved version of the deterministic analysis presented in this section through the modeling of cost growth and the stochastic comparison of UP and LS costs.

Cost Growth Modeling. Cost growth in this study is defined as the percentage variation of the original contract cost (total

Table 1. UP Projects: Cost Growth

Unit price projects									
No	Miles	Original cost ($)	Actual cost ($)	Cost growth (%)	No	Miles	Original cost ($)	Actual cost ($)	Cost growth (%)
1	0.66	521,217.55	506,672.13	−3	33	16.14	3,021,171.51	2,921,086.34	−3
2	0.77	648,208.71	638,066.90	−2	34	16.73	3,493,487.41	3,547,934.65	2
3	1.09	759,549.53	735,190.97	−3	35	17.17	4,955,711.06	4,790,431.51	−3
4	1.41	862,770.36	791,326.63	−8	36	17.47	3,529,617.70	3,475,221.87	−2
5	2.10	841,046.90	781,434.70	−7	37	17.63	4,883,159.85	4,581,603.56	−6
6	2.80	1,483,128.17	1,430,102.15	−4	38	18.46	3,802,688.89	3,479,142.77	−9
7	2.98	1,095,029.24	997,606.57	−9	39	19.88	4,178,282.93	4,017,351.92	−4
8	3.69	1,312,213.59	1,254,212.88	−4	40	20.00	5,112,998.18	5,183,132.95	1
9	3.96	966,283.04	953,705.99	−1	41	21.07	3,983,484.70	3,695,174.48	−7
10	4.38	896,919.51	777,644.27	−13	42	21.35	3,543,343.50	3,802,307.75	7
11	5.28	1,940,620.07	1,856,551.56	−4	43	22.09	4,199,999.48	3,826,661.87	−9
12	5.40	2,376,742.36	2,384,891.78	0	44	22.65	4,007,551.55	3,641,742.32	−9
13	6.38	1,463,327.52	1,424,598.85	−3	45	23.91	5,333,715.73	5,049,334.18	−5
14	6.74	1,656,116.65	1,659,137.45	0	46	24.04	4,933,423.90	4,649,829.70	−6
15	6.85	1,424,722.42	1,391,631.60	−2	47	25.16	6,085,725.81	5,952,142.50	−2
16	7.12	1,728,947.87	1,633,505.84	−6	48	26.44	4,054,389.57	3,749,496.73	−8
17	7.56	1,885,541.29	1,821,043.74	−3	49	27.22	4,037,986.70	3,846,478.33	−5
18	7.58	1,865,344.96	1,622,084.47	−13	50	27.70	8,076,272.17	7,667,982.38	−5
19	9.31	3,069,378.46	2,906,324.38	−5	51	28.96	4,962,037.99	4,756,358.08	−4
20	9.82	3,648,201.75	3,514,981.59	−4	52	31.10	6,196,552.04	5,669,933.59	−8
21	10.23	3,284,461.01	3,131,127.75	−5	53	34.84	7,717,553.92	7,576,664.71	−2
22	10.65	2,386,059.87	2,290,475.30	−4	54	35.28	7,289,392.69	7,140,234.06	−2
23	11.30	1,961,731.16	1,958,164.54	0	55	36.48	2,702,300.94	2,445,064.27	−10
24	11.42	5,490,800.16	5,404,113.44	−2	56	37.98	9,008,955.84	8,863,555.68	−2
25	11.43	1,619,320.78	1,566,710.74	−3	57	38.53	4,926,846.56	4,585,418.50	−7
26	11.73	2,539,890.92	2,497,433.75	−2	58	43.07	7,416,925.38	7,441,691.97	0
27	11.74	3,745,728.33	3,415,842.50	−9	59	43.15	6,785,910.43	7,183,335.97	6
28	11.78	4,039,522.34	4,226,492.10	5	60	45.48	8,823,739.00	7,800,286.81	−12
29	12.70	2,030,029.00	2,045,054.57	1	61	48.32	11,419,266.79	9,884,931.68	−13
30	12.79	6,621,886.77	6,552,074.16	−1	62	49.88	9,957,902.43	9,419,098.68	−5
31	15.22	2,097,226.82	1,978,887.02	−6	63	60.26	7,824,412.70	7,721,970.60	−1
32	15.67	2,706,520.06	2,796,761.63	3					

bid price from selected contractor) and the total actual cost at project completion, including change orders and cost adjustments [Cost Growth = (Actual Cost – Original Cost)/Original Cost]. The process to model cost growth for UP and LS projects consists of two steps: 1) the calculation of cost growth for each UP and LS project; and 2) the use of statistical inference methods to determine the type of probability distribution that best fits the cost growth values for UP and LS. Tables 1 and 2 show the cost growth values for all UP and LS projects, respectively, which correspond to the first part of the process.

In the second part of the process to model cost growth, the author used the chi-square goodness of fit statistical test to infer the most suitable probability distribution for each set of cost growth values. This statistical test was conducted using a statistical software package called @Risk that facilitates the calculation of the goodness of fit for several probability distributions. Using this statistical test, the authors found that cost growth values for UP projects follow a normal distribution with mean (μ) −3.9% and standard deviation (σ) 4.3%. This normal distribution with the frequency of cost growth values is illustrated in Figure 5.

On the other hand, the distribution of cost growth for LS projects seems to match an extreme value minimum distribution (Gumbel distribution) with alpha (location parameter: γ) and beta (scale parameter: β) values equal to −1.9% and 0.02 (mean = −3.2%; standard deviation = 2.8%), respectively. This probability distribution deals with data that have an extreme deviation from the mean (*15*). Extreme value distributions are commonly used in hydrology to estimate unusually large natural events such as 100-year flood peaks (*16*). The extreme value minimum distribution and distribution of growth values for LS projects is illustrated in Figure 6.

Table 2. LS Projects: Cost Growth

Lump sum projects				
Number	Miles	Awarded amount ($)	Final project cost ($)	Change (%)
1	0.98	219,504.00	213,917.00	−3
2	1.48	862,414.80	823,651.81	−4
3	2.73	664,503.82	636,000.00	−4
4	3.01	1,033,200.00	920,104.14	−11
5	4.98	1,688,000.00	1,606,582.22	−5
6	5.41	910,000.00	883,226.89	−3
7	6.23	1,558,478.00	1,525,500.59	−2
8	7.41	1,563,213.43	1,512,750.97	−3
9	7.56	1,795,728.27	1,760,006.21	−2
10	7.71	1,296,000.00	1,222,129.81	−6
11	7.75	1,298,500.00	1,319,720.21	2
12	9.32	2,509,778.30	2,358,043.58	−6
13	9.36	1,223,500.00	1,216,824.17	−1
14	9.54	2,888,000.00	2,634,048.09	−9
15	10.11	2,084,540.00	2,051,990.95	−2
16	12.15	1,947,842.63	1,891,898.60	−3
17	12.94	2,754,000.00	2,686,703.58	−2
18	13.26	3,276,000.00	3,298,885.66	1
19	13.69	5,895,900.00	5,727,577.20	−3
20	17.03	3,670,000.00	3,621,208.52	−1
21	19.39	3,757,548.78	3,684,887.35	−2
22	19.80	3,832,891.28	3,786,450.72	−1
23	26.28	6,951,000.00	6,782,907.00	−2

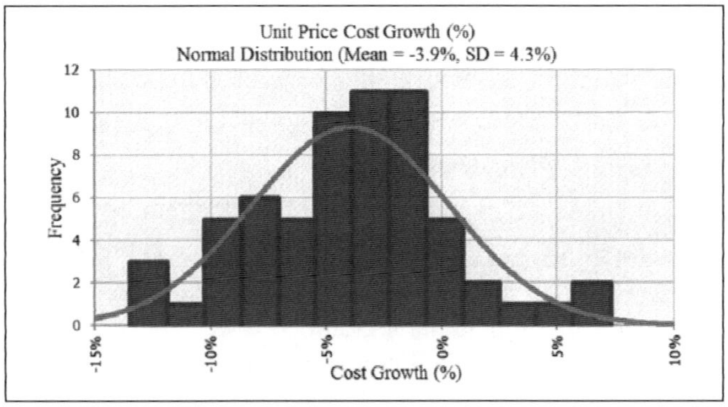

Figure 5. UP cost growth probability distribution.

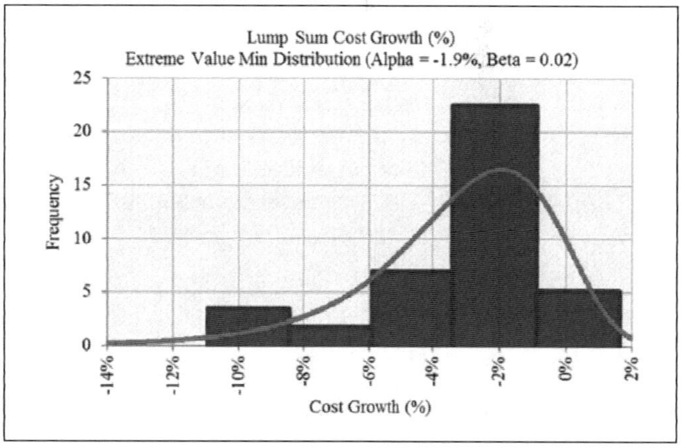

Figure 6. LS cost growth probability distribution.

Figure 7 helps to visualize the difference between the cost growth distributions for both types of projects by combining them into a single graph. A comparison between these two graphs shows similar negative mean values; −3.9% and −3.2% UP and LS respectively, indicating that actual final costs in both types of projects seem to be lower than original contract values. However, standard deviation values show that values in cost growth distribution for UP are more spread out than in the distribution for LS projects (standard deviation for UP = 4.3%; LS = 2.8%). That means LS projects have higher cost certainty—an important aspect in comparison of stochastic values presented in the following sections.

Stochastic Construction Cost Estimating Models: Development and Comparison. Stochastic cost estimates are produced by combining the deterministic and cost growth models. The combination of these models for UP and LS resurfacing projects was performed using Equations 1 and 2, respectively. What these equations are doing is to first estimate the deterministic cost for a given number of lane miles, and to then assess possible fluctuations in this value based on observed cost growth trends (cost growth model).

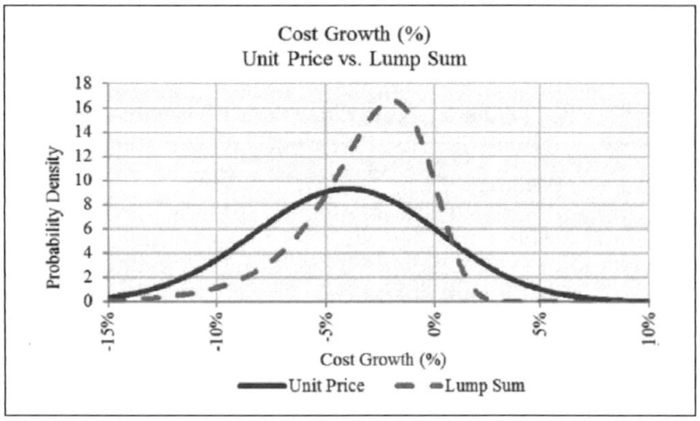

Figure 7. Cost variation: UP versus LS.

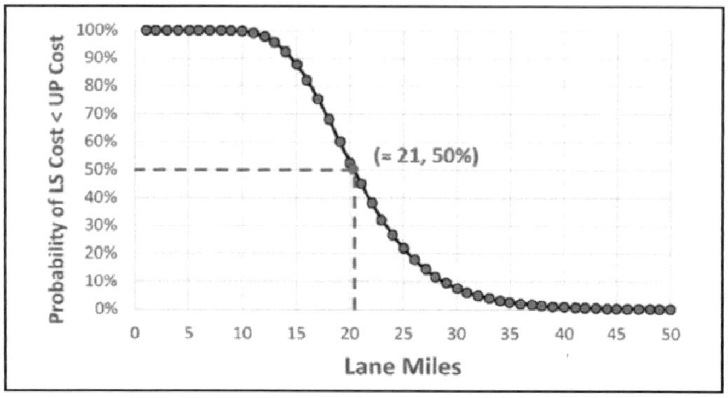

Figure 8. Probability UP will be less than LS.

$$SE_{(UP,x)} = DE_{(UP,x)} * (1 + N(-0.039; 0.043)) \quad (1)$$

where

x = lane miles,
$SE_{(UP,x)}$ = stochastic cost estimate for a UP project of x lane miles,
$DE_{(UP,x)}$ = deterministic cost estimate for a UP project of x lane miles, and
$N(-0.039; 0.043)$ = UP cost growth model→normal distribution (mean = −3.9%; SD = 4.3%).

$$SE_{(LS,x)} = DE_{(LS,x)} * (1 + EV(-0.019; 0.02)) \quad (2)$$

where

x = lane miles,
$SE_{(LS,x)}$ = stochastic cost estimate for an LS project of x lane miles,
$DE_{(UP,x)}$ = deterministic cost estimate for an LS project of x lane miles, and
$EV(-0.019; 0.02)$ = LS cost growth model→extreme value distribution (alpha = −1.9%; beta = 0.02).

Stochastic cost estimating models for UP and LS (Equations 1 and 2) were compared for different project sizes (number of lane miles) to determine the probability of UP costs being greater than LS costs as a function of the number of lane miles (ranged from 1 to 50 mi). The stochastic comparison was performed using a statistical software package (@Risk) and Monte Carlo techniques to simulate 10,000 pairs of random values for Equations 1 and 2 for each type of project and for each project size. In other words, 10,000 pairs of values (one UP and one LS value) were randomly generated and recorded for a resurfacing project of 1 lane mile; and the same was repeated 49 more times for different numbers of lane miles. The number of times UP costs exceeded LS costs was calculated and used to determine the probability of having greater costs if UP is used for each project size (number of times UP exceeded LS/10,000 * 100%). Figure 8 shows the probability for resurfacing projects ranging from 1 to 50 lane miles.

The stochastic model also shows that an LS approach tends to be more cost-efficient for resurfacing projects of less than 21 lane miles (approximately). As shown in Figure 8, the probability of UP costs exceeding LS costs is greater than 50% for resurfacing projects of less than 21 lane miles, which is the same break point obtained with the deterministic model. However, this graph also shows that for projects ranging from 10 to 35 lane miles there is still a chance of making the wrong decision. For resurfacing projects of less than 10 lane miles, LS is clearly the best option; contrary to projects for more than 35 mi in which UP is the best alternative. It is important to know the probability of making the wrong decision, but probably it is more important to know the consequences of making the wrong decision, which is not shown in Figure 8, and in some cases could be significant. Such information was collected during the simulation and provided to decision makers through the nomogram described in the next section.

Integration of Results

The nomogram presented in Figure 9 integrates the entire methodology proposed in this paper into a single two-dimensional graph that facilitates decision making under different confidence levels. The nomogram uses two inputs: number of lane miles; and desired confidence level set by decision makers. These two inputs produce three outputs: probability of having higher construction costs if UP instead of LS; the worst-case scenario if LS provisions are used; and the best-case scenario if LS provisions are used. The confidence level depends on the risk profile of the decision makers (*17*). The higher the confidence level, the larger the range between the worst- and best-case scenarios, and the higher the confidence of decision makers that the actual savings/costs of using LS on a given project will lay within this range.

If, for example, FDOT decides to use a 75% confidence level to decide whether to use a LS approach on a given project, that agency would be 75% sure that the actual cost implications of using LS provisions are between paying $11,000 of additional costs and saving $13,000 of extra costs (see Figure 9). The estimation of the worst- and best-case scenarios in dollars will make it easier to integrate the nomogram with FDOT's

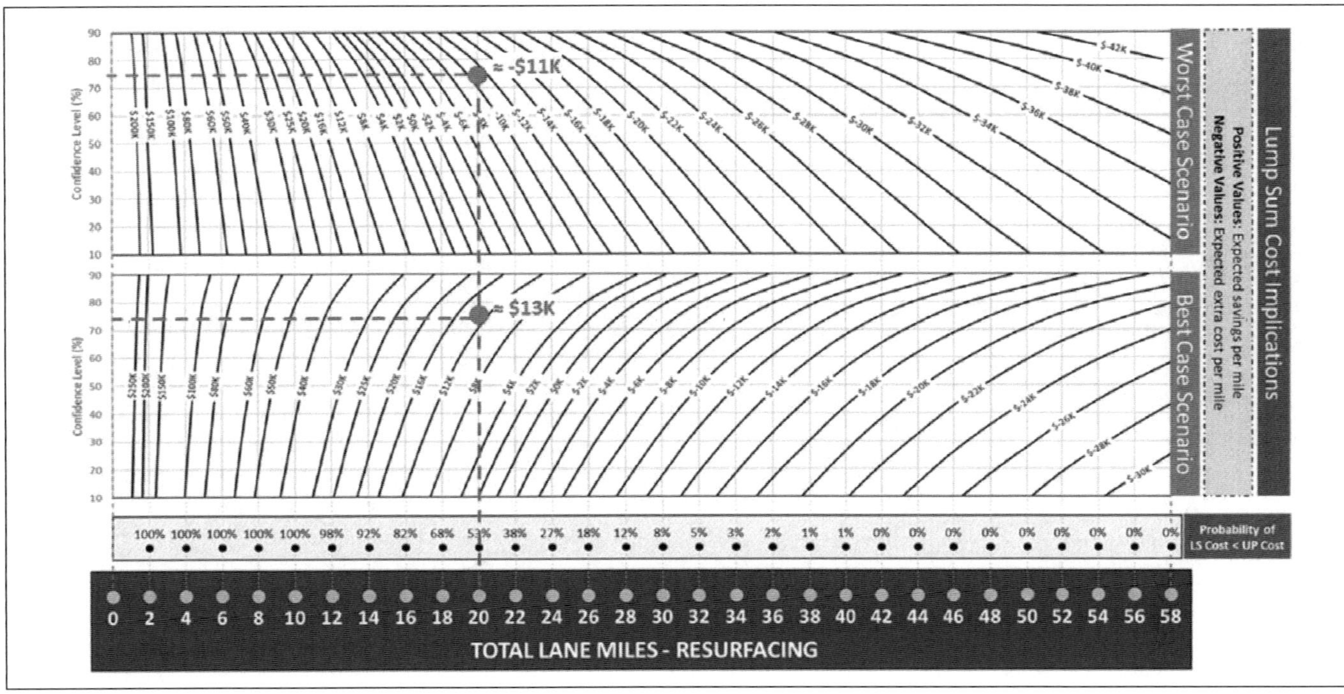

Figure 9. LS project selection: cost-based nomogram.

decision-making practices. If we add to the example in this paragraph that FDOT has estimated that the use of LS would represent an increase of $10,000 in design costs because of the higher design quality required in LS contracts, this additional cost could be integrated with the nomogram because they use the same units—dollars. In that case, using LS would be less attractive with a potential saving of $3,000, in the best-case scenario, or extra costs up to $11,000.

Conclusion and Recommendations

Despite the increasing use of alternative project delivery methods, traditional DBB contracting techniques are still the most common contracting method used by STAs, and any research efforts toward their improvement are expected to have a great impact on STAs' construction programs. This paper summarizes the research efforts made to develop a stochastic cost-based decision-making tool to identify suitable projects for DBB-LS for FDOT. The proposed methodology compares the UP and LS stochastic estimates for candidate projects and provides decision makers with the probability of UP costs exceeding LS costs, as well as an estimate of the cost implication of using LS.

The development and implementation of the decision-making tool was illustrated using bid data from 86 resurfacing projects completed by FDOT between 2015 and 2017 (63 UP and 23 LS). Several tools/techniques were used to process this data into cost growth models and deterministic/stochastic cost estimating models for UP and LS. The primary contribution of this study is a nomogram that provides decision makers with three valuable outputs: the probability of having higher construction costs if UP is used instead of LS; the worst-case scenario if LS provisions are used; and the best-case scenario if LS provisions are used.

Future research might consider design and project administration costs as criteria in the selection of projects for LS as well as a thorough cost–benefit analysis to gain a better understanding of the risks associated with the potential savings offered by LS provisions. Likewise, the authors recommend expanding this study to other types of work besides resurfacing.

References

1. Gransberg, D. D., and J. S. Shane. *NCHRP Synthesis 402: Construction Manager-at-Risk Project Delivery for Highway Programs.* Transportation Research Board of the National Academies, Washington, D.C., 2010.
2. Molenaar, K. R, C. Harper, and D. Tran. *Guidebook for selecting project delivery methods and alternative contracting strategies.* University of Colorado, Colorado, 2012.
3. Anderson, S. D., and I. Damnjanovic. *NCHRP Synthesis 379: Selection and Evaluation of Alternative Contracting Methods to Accelerate Project Completion.* Transportation Research Board of the National Academies, Washington, D.C., 2008.
4. Walewski, J., G. E. Gibson, Jr., and J. Jasper. *Project Delivery Methods and Contracting Approaches Available for Implementation by the Texas Department of Transportation.* FHWA/TX-0-2129-1. Center for Transportation Research, Austin, Texas, 2001.
5. Miller, J. B. Engineering systems integration for civil infrastructure projects. *Journal of Management in Engineering*, Vol. 13, No. 5, 1997, pp. 61–69.

6. Scott, S., and K. Mitchell. *Alternative Payment and Progress Reporting Methods*. Federal Highway Administration, Trauner Consulting Services, Inc. https://www.fhwa.dot.gov/programadmin/contracts/etgpayment.cfm. Accessed July 26, 2017.
7. Ellis, R. D., J. Pyeon, Z. J. Herbsman, E. Minchin, and K. Molenaar. *Evaluation of Alternative Contracting Techniques on FDOT Construction Projects*. Final Report. Florida Department of Transportation, Department of Civil and Coastal Engineering, University of Florida, 2007.
8. Minchin, R. E., Jr., A. R. Chini, and M. E. Rinker. *Alternative Contracting Research*. Final Report. FDOT Contract Number BDV31-977-40, Florida Department of Transportation, 2016.
9. Florida Department of Transportation. *Plans Preparation Manual*. Florida Department of Transportation, 2017.
10. Transportation and Public Facilities. *Lump Sum Project Guidelines*. Alaska Department of Transportation and Public Facilities, 2003.
11. Trauner Consulting Services. *Alternative Procurement and Contracting Methods, Innovative Procurement Practices*. Contract no. 53A0104. California Department of Transportation, 2007.
12. Florida Department of Transportation. *Contracts Administration – Lettings & Project Information*. Florida Department of Transportation, 2017. http://www.fdot.gov/contracts/Lettings/Letting_Project_Info.shtm. Accessed August 1, 2017.
13. McGee, Sr, Hugh W. *NCHRP Synthesis 417: Geometric Design Practices for Resurfacing, Restoration, and Rehabilitation*. Transportation Research Board of the National Academies, Washington, D.C., 2011.
14. Rueda-Benavides, J. A. *Indefinite Delivery/Indefinite Quantity Project Selection Framework Using Stochastic Techniques*. Doctoral dissertations. Iowa State University, 2016.
15. Gumbel, E. *Statistics of Extremes*. Columbia University Press, New York, 1958.
16. Abarbanel, H., S. Koonin, H. Levine, G. MacDonald, and O. Rothaus. *Statistics of Extreme Events with Application to Climate*. Mitre Corp, McLean, Va., Jason Program Office, 1992.
17. Chelst, K., and Y. B. Canbolat. *Value-Added Decision Making for Managers*. CRC Press, Boca Raton, 2011.

The Standing Committee on Project Delivery Methods (AFH15) peer-reviewed this paper (18-00498).

Article

TRR
JOURNAL OF THE TRANSPORTATION RESEARCH BOARD

Transportation Research Record
2018, Vol. 2672(26) 21–27

DOI: 10.1177/0361198118758030
journals.sagepub.com/home/trr
SAGE

Life-Cycle Cost Adjustment Factors in Alternate Design/Alternative Bid Pavement Bids: Added Value or Added Controversy?

Douglas D. Gransberg[1], Ilker Karaca[1], and Ashley Buss[1]

Abstract

Alternative design/alternative bids (ADAB) provides a mechanism for the asphalt and concrete paving industries to compete for the same paving project. It operates on the principle of the market pricing of each material determining which is most economical when the bids are opened, rather than selecting the pavement type during design based on a life-cycle cost analysis (LCCA). This paper reviews including LCC-based bid adjustment factors in the ADAB award decision. Data are from a survey that received responses from 40 U.S. Departments of Transportation (DOT) and the Canadian province of Ontario, and a content analysis of 55 ADAB project outcomes in 13 U.S. states and three Canadian provinces. Seven algorithms in use to calculate an ADAB bid adjustment factor were found, and six U.S. DOTs that award ADAB projects without an adjustment factor. The paper finds that the adjustment factor formula rarely influences the award decision and, generally, the pavement type with the lowest bid cost wins with or without the adjustment factor. The paper models the ADAB process in financial terms as an exercisable commodity option that accrues value from the differential rates of volatility between asphalt and concrete. It concludes that an LCC-based bid adjustment factor complicates the award process, creating potential for controversy over what the factor inputs are, and does not add value over bidding the pavement types head to head and awarding to the low bidder. The ADAB process increases the number of bidders and reduces unit bid prices for both pavement types.

Selecting the pavement type for a highway construction project is a decision that transcends the technical design process. It has become increasingly controversial over the past several decades as the asphalt and concrete paving industries fight for market share by trying to influence the decisions made by public officials on fundamental highway design. The issue is made more contentious by the persistent perception by both engineers and non-engineers that is expressed succinctly by a quote from Suvo and Stonecypher: "Concrete roads are highly durable and more environmentally friendly as compared to asphalt roads. However, asphalt paving costs far less than concrete paving" (*1*). The fight spilled out of the technical realm and into the financial realm when the FHWA promoted the use of life-cycle cost analysis (LCCA) as the means to determine the most economical pavement type during the design process (*2*). The quest for the most economical solution eventually led to the development of alternative design/alternative bids (ADAB) as a means of providing a mechanism to allow both industries to compete for the same project (*3*). However, even when both materials are available as acceptable options, the controversy over the means for selecting the winning bidder remains because the generally accepted approach is to develop an LCC adjustment factor that is applied to the bid price for the asphalt, and sometimes also the concrete bid, to account for the difference in service lives and maintenance costs.

The objective of this paper is to detail the state of the practice regarding ADAB and LCC adjustment factors, and then discuss the technical and financial advantages and disadvantages of the LCC adjustment factor in ADAB bidding. The information provided comes from a 2016 NCHRP study on the topic which drew survey responses and case studies from 40 U.S. state Departments of Transportation (DOTs) and the Ministry of Transportation of the Canadian province of Ontario (MTO). The study identified seven different formulae

[1]Department of Civil, Construction, and Environmental Engineering, Iowa State University, Ames, IA

Corresponding Author:
Address correspondence to Douglas D. Gransberg: dgran@iastate.edu

for calculating the LCC adjustment factor and identified six DOTs that bid hot-mix asphalt (HMA) and Portland cement concrete (PCC) pavement types head to head without an LCC adjustment factor. Lastly, the paper posits that although the philosophy of LCCA-based pavement design decisions is certainly intuitive, the practice is far from standardized, as seen from the seven different LCC adjustment algorithms, and that the volume of LCCA research indicates that LCC is highly sensitive to the input assumptions made for variables, like the discount rate, the period of analysis, and the residual (salvage) value for the pavement type with the longer life (*2, 4, 5*). These facts make one question the utility of the LCC adjustment factor itself in the ADAB bidding process, considering the well-documented continuing controversy over appropriate input values (*4–6*).

Background

NCHRP Synthesis 499: *Alternate Design/Alternate Bid Process for Pavement-Type Selection* concluded the following about ADAB practices:

> Alternate Design/Alternate Bid (ADAB) allows the pavement-type selection decision to be made as part of the procurement process by permitting contractors to bid their preferred alternative using real-time market pricing for the paving materials. This synthesis found that many ADAB projects documented an increased number of bidders on a given paving project by allowing both the asphalt and concrete paving industries to compete. ADAB projects also document a general trend of overall bid unit price reduction for both pavement types. (*3*)

FHWA *Technical Advisory T 5040.39: Use of Alternate Bidding for Pavement-Type Selection* (*7*) defines ADAB as a procurement in which two or more equivalent pavement designs are furnished, allowing the marketplace to determine the most economic pavement type based on real-time pricing for both alternatives at the time of letting. The Technical Advisory advises that ADAB is "a suitable approach for determining pavement type when engineering and economic analysis does not indicate a clear choice between different pavement designs" (*7*) Synthesis 499 also found that many agencies that use ADAB have established a criterion for the relative values of asphalt and concrete pavement LCCs, such as being within a given percentage of each other, as the trigger for implementing ADAB on a given project (*3*). The synthesis found that the cost of preparing two sets of equivalent designs was trivial. In fact, the Indiana DOT calculated a saving of $3.8 million on nine ADAB projects, which yields a benefit cost ratio of 11.4 for the additional fee paid to its design consultants for the extra pavement design (*8*).

Lastly, Synthesis 499 found that "implementing ADAB contracts can lead to increased competition and reduced pavement material costs" (*3*). The Missouri DOT reported an average increase of two more bidders per letting, and a 5% to 8% decrease in price in asphalt and concrete paving prices in over 187 lettings. The Indiana, Louisiana, and Michigan DOTs all reported a decrease in pricing on their ADAB projects when compared with conventional paving projects. The Portland Cement Association maintains ADAB "gives the contractor a choice to bid on either a concrete or an asphalt option, thereby increasing the number of bidders on each job and enhancing competition" (*9*).

Impact of Construction Material Price Volatility

The difference between classical LCCA and the development of ADAB LCC adjustment factors has to do with the timing of the pricing data. FHWA provides guidance for LCCA that is intended to compare the total whole life costs of design alternatives to make the pavement-type selection decision (*2*). The FHWA LCCA primer maintains that construction costs are only a part of the analysis and it provides a method for comparing the life-cycle costs and benefits of possible pavement design alternatives (*10*). Typical pavement includes costs for design, construction, maintenance, rehabilitation, and user costs over a period of analysis spanning at least one major rehabilitation (*11*). However, this analysis is conducted using estimate prices for costs that will all occur several years in the future after the project is awarded. Therefore, estimates for material price escalation must also be made, and this is when volatility in the construction materials market can affect actual costs.

The term "price escalation" denotes an inherent bias to inflate all future costs as a conservative measure to keep the project's budget sufficient to allow an award as scheduled. This philosophy unintentionally eliminates the ability to accrue savings as a result of deflation in one commodity versus its competing alternate. Figure 1 shows the differential changes in liquid asphalt and Portland cement commodity prices over the period of January 2004 to June 2017 (*12*). The figure shows that during the period from June 1, 2008 to December 1, 2008 that asphalt's commodity price experienced a swing of 16% (+11% in June to −5% in December). In the same period, cement's volatility was only about 2%. In the first 5 months of 2017, the asphalt volatility ran from +8% to −5% with cement running from +3% to −1%.

Figure 2 shows the Federal Reserve Bank of Saint Louis' asphalt and Portland cement indexes over the same period (*12*). Although the two metrics are not calibrated to one another, they have been graphed in juxtaposition to provide a visual representation of how they have moved with respect to each other over the same period as the commodity prices shown in Figure 1. Figure 2 appears to indicate a slight trend to the two indexes changing in opposition to each other. In other words, when asphalt goes up, cement goes down, and vice versa. The information provided in the figures leads to two conclusions. First, making pavement-type selection decisions with LCCA based on pricing assumptions established years before letting, and escalated at some fixed discount rate to the future, is an exercise in futility. The

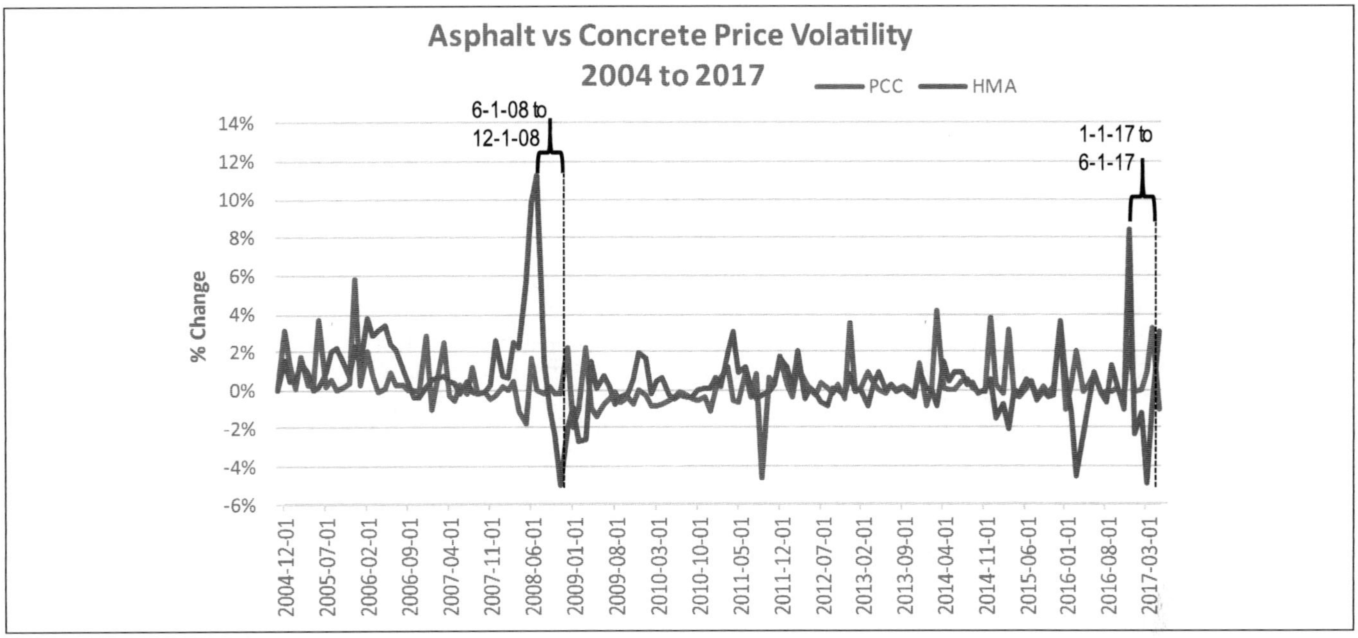

Figure 1. Pavement type volatility (*12*).

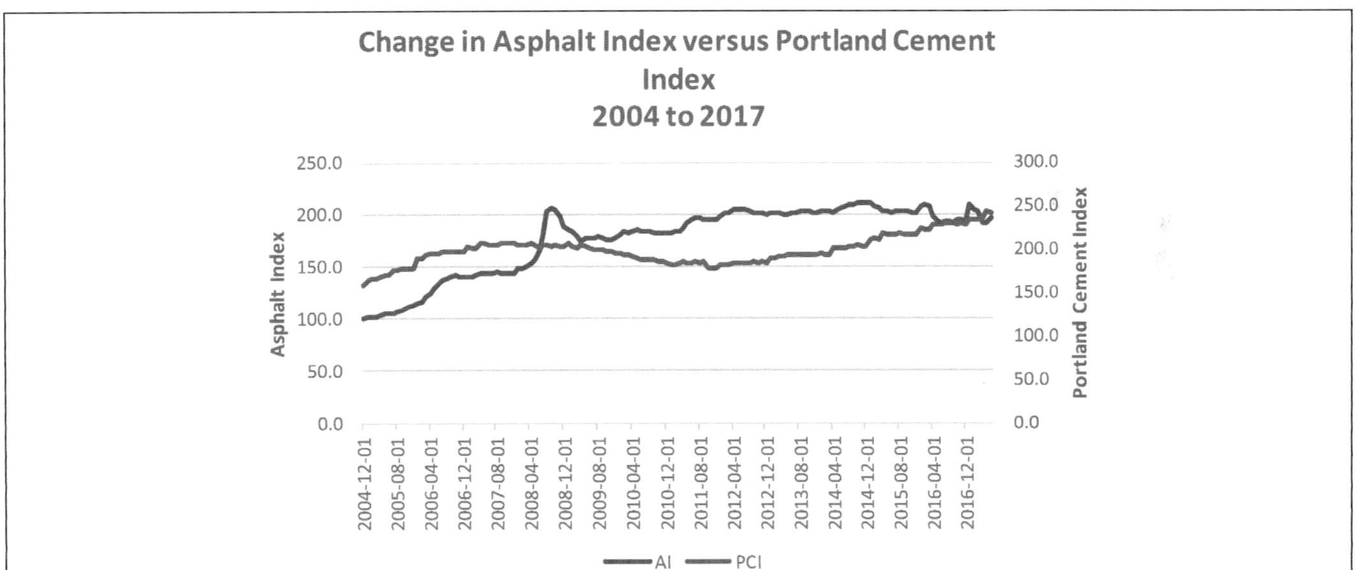

Figure 2. Relative change in asphalt and Portland cement indexes (*12*).

conventional perception that "asphalt paving costs far less than concrete" (*1*) is faulty. Second, by selecting a pavement type before the day of letting, the agency is giving up the opportunity for cost savings accrued from a favorable swing in the relative prices of the two pavement types in the commodities market.

Whereas vast volumes of forward and futures contracts are traded daily on commodity exchanges to profit from the types of volatilities described above, similar hedging strategies are not common in construction projects. For instance, the daily trading volume for West Texas Intermediate futures, which could be used to hedge exposures to asphalt prices in a pavement project, is over $50 billion (*13*).

The price volatility of different construction inputs under varying pavement-type designs creates the so-called option value, as it is referred to in the finance literature (*14*). In fact, deferring the timing of the pavement-type selection decision to test the market prices of competing design alternatives would be described as "deferment (learning) real options" (*15*). Thus, once the value embedded in the ADAB practices is recognized as option values (i.e., potential cost savings from allowing alternative design proposals to compete based on

Table 1. LCC-Based Bid Adjustment Factor Summary (*3*)

Name	Formula	Highway agencies
A + C bidding; C applied to HMA only	HMA bid = HMA contract bid amount + NPW of future HMA rehab PCC bid = PCC contract bid amount	Kansas, Montana, Oklahoma
A + C bidding; C applied to both	HMA bid = HMA contract bid amount + NPW of future HMA rehab PCC bid = PCC contract bid amount + NPW of future PCC rehab	Indiana, Kentucky
A + B + C bidding	HMA bid = HMA contract bid amount + value of time + NPW of future HMA rehab PCC bid = PCC contract bid amount + value of time + NPW of future PCC rehab	Kentucky, Louisiana
A + C + lane rental	HMA EUAC = HMA contract bid amount + lane rental + future HMA maintenance costs (capital recovery factor at OMB discount rate for 26 years) PCC EUAC = PCC contract bid amount + lane rental + future PCC maintenance costs (capital recovery factor at OMB discount rate for 26 years)	Michigan
Adjustment factor	Adjustment factor = NPW future asphalt rehab – PW future concrete rehab Low bidder = lower of (PCC bid price) versus (HMA bid price + adjustment factor) (Assuming asphalt has higher NPW M&R costs)	Missouri
LCC advantage	Low bidder = lower of (PCC bid price + NPW future concrete M&R) versus (HMA bid price + NPW future HMA M&R)	Ontario
A – D (alternative differential) bidding	Adjustment factor = fixed value set by DOT for each project Low bidder = lower of (PCC bid price- adjustment factor) versus HMA bid price (Assuming asphalt has higher NPW M&R costs)	Iowa
No adjustment	Low bidder = lower of PCC bid price versus HMA bid price	Alabama, Arkansas, Florida, North Carolina, Ohio, West Virginia

Note: NPW = Net Present Worth; EUAC = Equivalent Uniform Annual Cost; OMB = US Office of Management and Budget; M&R = Maintenance and Repair.

market conditions), the decision of whether to employ ADAB is merely a decision on the timing of when the agency's option is exercised. That is, selecting the final pavement type during design exercises the agency's option and forgoes the possibility of accruing benefits by exercising its option at letting by using ADAB. According to one author: "Bidding pavements 'head-to-head' allows the open market to determine what is constructed … not outdated assumptions made during the [pavement-type] evaluation/selection process years before letting" (*16*). The Missouri DOT made the decision to employ ADAB on all projects that exceeded 7,500 yd^2 of paved surface in 1996 (*17*). By doing so the Missouri DOT made the business decision to exercise its pavement material option as late as practical and, as will be discussed later in the paper, accrued substantial benefits as a result.

Methodology

The information contained in this paper was extracted from NCHRP Synthesis 499 (*3*). That broader effort collected survey data from 40 U.S. state DOTs (an 80% response rate), case studies from five U.S. state DOTs and the MTO in Canada, as well as a content analysis of the research literature on the topic. This paper goes on to prepare a detailed analysis of the LCC adjustment factor's impact on the pavement-type selection for a typical ADAB project to establish a conclusion on the utility of those factors in the overall process. The paper answers the following research question:

> *Does the inclusion of a LCC adjustment factor in the algorithm for determining the most economical bid price add value to the procedure?*

Table 1 shows the options found in NCHRP Synthesis 499 (*3*) that are used by DOTs to calculate a rational LCC adjustment factor. It is important to note that there seems to be no single uniform method for accounting for the difference in pavement-type service lives, post-construction operations and maintenance costs, or the amount of disruption the public will experience during construction. In addition to the various adjustment factor options, the survey found that six states apply ADAB without adjusting the bid pricing and rely on direct competition between the pavement types.

One of the major differences that can be observed in Table 1 is a difference in philosophy as to just how the LCC

Table 2. LCC Adjustment Factor Outcome Analysis

Bid	Alt	Initial bid	NPW future costs	A + C bidding; C applied to HMA only	A + C bidding; C applied to both	Adjustment factor	LCC advantage	A – D (alternative differential) bidding	No adjustment
A	HMA	29,072,406	827,376	29,072,406	29,899,782	29,899,782	29,899,782	28,245,030	29,072,406
B	PCC	25,747,900	356,240	25,747,900	26,104,140	26,104,140	26,104,140	25,391,660	25,747,900
C	PCC	27,978,568	356,240	27,978,568	28,334,808	28,334,808	28,334,808	27,622,328	27,978,568
D	PCC	29,999,350	356,240	29,999,350	30,355,590	30,355,590	30,355,590	29,643,110	29,999,350
E	PCC	43,419,000	356,240	43,419,000	43,775,240	43,775,240	43,775,240	43,062,760	43,419,000

Table 3. Summary of ADAB Project Outcomes Found in the Literature

Outcome		Outcome		Metric
Total ADAB projects	55	Total bids submitted	313	5.7 bids/project
Total HMA bids submitted	137	No. projects when HMA won	18	33%
Total PCC bids submitted	176	No. projects when PCC won	37	67%
No. projects with no HMA bids submitted	2	Total bids received when HMA won	80	26%
No. projects with no PCC bids submitted	4	Total bids received when PCC won	233	74%

adjustment should be applied. Five DOTs (Iowa, Kansas, Missouri, Montana, and Oklahoma) only apply the factor to the HMA option, whereas the rest apply it to both options. Failing to apply the factor to both options may induce an unintentional bias, as one option is being compared on the cost of its entire life against the other in which only the construction cost is considered. However, it is not the intention of this paper to critique the academic validity of each method. It will merely conduct an analysis of outcomes observed when each method is applied to the same project.

Analysis and Discussion

The analysis is straightforward. The details of an ADAB project that was found in the Special Experimental Project No. 14 (SEP-14) report series was extracted, and the LCC-adjusted bid price was calculated for the same set of bids (*18*). The project was the Shelbyville By-Pass Project in 2006 for construction of an overlay and widening of a four-lane concrete pavement in Shelby County, Kentucky (*18*). As this particular project did not contain actual data for value of time or lane rental, the A + B + C bidding and A + C + lane rental options were not included in the analysis. Table 2 contains the results of the analysis, which are quite consistent. Regardless of the presence or composition of an LCC adjustment factor, the low bidder was the same as shown by the tinted line in Table 2 for Bidder B.

The reader must keep in mind that Table 2 applies only to the specific project and the market conditions at the time of the bid. The relative unit pricing of HMA and PCC pavements move independently of each other and each material has its own level of price volatility. Thus, the results cannot be generalized. However, a report issued by the Missouri DOT found that out of 187 ADAB projects completed through 2010, the LCC adjustment factor only reversed the decision four times: about 2% of the projects (*19*). If the experience in Missouri is representative of that found in the nation, then a result like that shown in Table 2 would be more often the rule and, regardless of the composition of the LCC adjustment factor, it would not make a difference 98% of the time. Thus, it was concluded that, at least in Missouri, competing HMA and PCC head to head can be successfully implemented purely on the basis of lowest bid if the Missouri DOT chose to do so.

Table 3 attempts to put a broader representation on the Table 2 outcome by furnishing the details for 55 ADAB projects in 13 U.S. states and three Canadian provinces found in the literature. For each project, the number of bids submitted for each pavement type, the low HMA bid, the low PCC bid, the percentage difference between the two bids, and the winning pavement type were tabulated. Not all the bids were adjusted by an LCC-based factor because, as stated above, several DOTs do not use an adjustment factor and, furthermore, it is known that some agencies, like the Missouri DOT, that use a factor do not adjust all the pavement project bids (*17, 19*).

It can be seen in Table 3 that a total of 313 individual bids were attracted by ADAB projects, which translates to an average of 5.7 bidders per project. According to the literature, the average number of conventional project bids ranges from 3.8 to 5.1 bidders per project (*20, 21*). Therefore, it appears that the sample confirms the individual state results in the literature by exceeding the upper bound of the range. Next, the sample shows that 44% of the bids were for HMA and 56% were for PCC. Although not equal, the difference between the two pavement types is not large, supporting the idea that implementing ADAB does indeed give both industries an opportunity to compete. This trend is reinforced by the fact that only six of 55 projects (11%) received bids for only one pavement type.

The outcomes found in Table 3 show that PCC pavement bids won 67% of the time. The unscientific sampling procedure makes it difficult to draw authoritative conclusions based on the observed differences, but when combined with the results found in Missouri and the Table 2 demonstration of the efficacy of the various LCC factors in use, it does suggest that the inclusion of an adjustment factor to ensure a "level playing field" for PCC may not be needed. As the LCC adjustment factor is calculated before advertising an ADAB project, the volatility issues discussed in the previous section come into play as the agency "exercises its option" by calculating the factor based on unit prices current at the time the bid package is assembled. If the relative price swings shown in Figure 1 are possible, then it is not illogical to posit that the factor could be calculated when a given commodity price is high, and 2 to 3 months later when the bids are opened that the trend has reversed itself, making the LCC adjustment factor an inaccurate representation of the relative LCC.

Furthermore, research has repeatedly demonstrated that classic LCCA is highly sensitive to the input values that are based on professional judgment or assumptions, specifically the discount rate, the analysis period, and salvage value (*4*). Stone concludes that "the two sides had completely opposite views on the inclusion of price adjustment clauses when calculating initial construction costs and material specific escalation rates" (*22*). Thus, the composition of the adjustment factor itself creates controversy, further muddying the outcomes to the point that seven different "approved" versions of the factor are now in use to account for the same issue. When the above discussion is combined with the fact that six states do not use bid adjustment factors in their ADAB programs, it appears that the utility of the LCC-based bid adjustment factor is questionable.

Conclusion

The answer to the research question posed in the section on methodology is developed in the following manner:

- Portland cement and liquid asphalt have different rates of volatility and may even change in opposition to each other. Those rates have been extreme, a 16% swing in 5 months. When the pavement-type selection decision is viewed as a business decision in purely financial terms, it can be modeled as exercising an option in the commodities market, which according to standard financial theories becomes more valuable as volatility increases. Therefore, ADAB as means to "exercise the option" at the last possible moment creates value for money in and of itself.
- The controversy between the asphalt and concrete paving industries with regard to the proper calculation of pavement LCCs is both intense and long standing. The result is that at least seven different algorithms are in use to adjust bid prices in ADAB procurements. Some of the methods do not include all the elements of the pavement LCCA prescribed by the FHWA. Thus, it can be concluded that the current suite of different adjustment factors is the product of negotiation (*16, 17, 21, 23*), not hard, scientific engineering economics that should have only a single, theoretically correct, method for solving the problem used by all (*2*).
- The example shown in this paper and the experience of the Missouri DOT indicate that even when an adjustment factor is applied, it rarely changes the award from the lowest priced pavement type. In addition, the content analysis of ADAB project award outcomes disproves the perception that asphalt is always the lowest price option in that PCC won 67% of the 55 ADAB projects reported in the literature.
- Six U.S. states currently let ADAB projects with no bid adjustment factors. Of those, the West Virginia DOT reported a saving of $16.4 million on six ADAB projects (*9*). The Ohio DOT, based on implementing alternate bidding on 10 projects, reported savings of $58 million, with the average winning bid price for ADAB projects 15% below the engineer's estimate in 2009 (*23*).

Therefore, the research question's answer is plain. LCC-based bid adjustment factors do not add value to the ADAB process. They add controversy and may unnecessarily complicate a relatively simple process: opening the bids on bid day and awarding to the lowest bidder regardless of pavement type (*24*). The potential for industry opposition may have a chilling effect that has stopped some agencies from accruing the value inherent to the differential levels of volatility of the two commodities available by treating ADAB as an exercisable option in a business rather than as an engineering decision.

Acknowledgments

The authors gratefully acknowledge both the financial and technical support of the NCHRP Synthesis 499 program and panel members.

References

1. Suvo, A., and L. Stonecypher. *Concrete Roads versus Asphalt Roads*. Bright Hub Engineering, Troy, NY, 2010. http://www.brighthubengineering.com/concrete.../45858-concrete-roads-vs-asphalt-roads/. Accessed July 18, 2017.
2. Walls, J., and M. R. Smith. *Life-Cycle Cost Analysis in Pavement Design —Interim Technical Bulletin*. FHWA-SA-98-079. FHWA, U.S. Department of Transportation, 1998.

3. Gransberg, D. D, A. Buss, I., Karaca, and M. C. Loulakis. *Alternate Design/Alternate Bid Process for Pavement-Type Selection*. NCHRP Synthesis 499. Transportation Research Board of the National Academies, Washington, D.C., 2017, p.3.
4. Gransberg, D. D., and E. Scheepbouwer. Infrastructure Asset Life Cycle Cost Analysis Issues. *2010 AACE International Transactions*, Atlanta, Ga., June 2010, pp. CSC.03.01–CSC.03.8.
5. Jeong, D., and S. Abdollahipour. *Alternate Bidding Strategies for Asphalt and Concrete Pavement Projects Utilizing Life Cycle Cost Analysis (LCCA)*. Report OTCREOS10.1-20-F. Oklahoma Transportation Center, Midwest City, Okla., 2012.
6. Pittenger, D. M., D. D. Gransberg, M. Zaman, and C. Riemer. Stochastic Life Cycle Cost Analysis for Pavement Preservation Treatments. *Transportation Research Record: Journal of the Transportation Research Board*, 2012. 2292: 45–51.
7. FHWA. *Technical Advisory T 5040.39: Use of Alternate Bidding for Pavement-Type Selection*. FHWA, U.S. Department of Transportation, Dec. 2012, p.2.
8. Duncan, T. L., and D. B. Holtz. Alternate Bidding History and Requirements. *Presentation 11th International Conference on Concrete Pavements, San Antonio, Texas,* 2012, p. 13. http://docs.lib.purdue.edu/cgi/viewcontent.cgi?article=1231&context=roadschool. Accessed May 20, 2016.
9. Portland Cement Association (PCA). State Save Millions in Paving Costs with Alternate Design and Bidding. White Paper. http://www.cement.org/. Accessed July 24, 2017.
10. FHWA. *Life-Cycle Cost Analysis Primer*. U.S. Department of Transportation Federal Highway Administration Office of Asset Management, Aug. 2002. http://www.fhwa.dot.gov/asset/lcca/010621.pdf. Accessed July 24, 2017.
11. Hallin, J. P., S. Sadasivam, J. Mallela, D. K. Hein, M. I. Darter, and H. L. Von Quintus. *NCHRP Report 703: Guide for Pavement-Type Selection*. Transportation Research Board of the National Academies, Washington, D.C., 2011.
12. Federal Reserve Bank of Saint Louis. Economic Research, FRED Economic Data. https://fred.stlouisfed.org/. Accessed July 26, 2017.
13. CME Group. NYMEX WTI Light Sweet Crude Oil Futures Briefing. 2017. http://www.cmegroup.com/trading/why-futures/welcome-to-nymex-wti-light-sweet-crude-oil-futures.html. Accessed Jul. 26, 2017.
14. Luehrman, T. A. Strategy as a Portfolio of Real Options. *Harvard Business Review*, Vol. 76, 1998, pp. 89–101.
15. Amram, M., and N. Kulatilaka. *Real Options: Managing Strategic Investment in an Uncertain World*. Oxford University Press, New York, 1998, p.112.
16. Lenz, R. TxDOT's Alternate Pavement Design and Bid Policy. Presented at 2010 Texas Asphalt Paving Association Annual Meeting, Austin, Tex., 2010, p. 2.
17. Roark, N. Alternative Pavement Bidding in Missouri, Unpublished presentation to the 2011 Virginia Concrete Conference, Richmond, Va., March 3, 2011.
18. Newman, O. G. *Special Experimental Project No. 14 (SEP-14) Evaluation of Alternate Pavement Bidding*. Kentucky Transportation Cabinet (KYTC), Ky., 2008. http://www.fhwa.dot.gov/programadmin/contracts/sep14kyeval.cfm. Accessed July 26, 2017.
19. Ahlvers, D. MoDOT Alternate Pavement Approach, Unpublished presentation to the 2010 North Central Asphalt User/Producer Group, Overland Park, Kans., Feb. 3, 2010. https://engineering.purdue.edu/~ncaupg/Activities/2010/Presentation/Ahlvers%20MODOT%20Alternate%20Pavement%20Approach.pdf. Accessed May 1, 2016.
20. Indiana Department of Transportation (INDOT). *Initial Evaluation Report, SEP-14, Alternate Bids on Pavement Type Contract No. R-30106 & R-31948, US 31*. INDOT, Indianapolis, Ind., 2009. http://www.fhwa.dot.gov/programadmin/contracts/sep14in2009eval.pdf. Accessed May 20, 2016.
21. Mikesell, L. A. Alternative Pavement Bidding, SEP-14 Project Interim Report. MDOT, Lansing, MI, Nov. 5, 2012. http://www.fhwa.dot.gov/programadmin/contracts/sep14list.cfm. Accessed Jan. 28, 2016.
22. Stone, M. L. C. *Development of Unit Cost Estimating Models with Respect to Scale Economies and Material Price Volatility for Use in Probabilistic Life Cycle Cost Analyses*. PhD dissertation. University of Alabama, Tuscaloosa, 2013.
23. Ohio Department of Transportation (ODOT). *Final Report Special Experimental Project No. 14 for CLAIMAD-70*. ODOT, Columbus, Ohio, Dec. 2004. http://www.fhwa.dot.gov/programadmin/contracts/sep14oh2004b.cfm. Accessed April 11, 2016.
24. Gransberg, D. D., and C. Riemer. Impact of Inaccurate Engineer's Estimated Quantities on Unit Price Contracts. *Journal of Construction Engineering and Management*, Vol. 135, No. 11, 2009, pp. 1138–1145.

The Standing Committee on Construction (AFH00) peer-reviewed this paper (18-01133).

Article

TRR
JOURNAL OF THE TRANSPORTATION RESEARCH BOARD

Transportation Research Record
2018, Vol. 2672(26) 28–39

Article reuse guidelines:
sagepub.com/journals-permissions
DOI: 10.1177/0361198118758690
journals.sagepub.com/home/trr
SAGE

Development of an Automated Remote Asphalt Paving Quality Control System

Sicong Zhu[1], Xiangdi Li[1], Haoyang Wang[2,3], and Dongxiao Yu[4]

Abstract

The asphalt pavement construction monitoring system presented in this paper covers two main asphalt pavement construction processes: hot mix asphalt (HMA) plant production and the laydown and compaction operation. The main architecture of the monitoring system, the development of the required hardware and the software structure, and the implementation process are presented in this paper. The system is able to perform automatic data collection and data transmission, and can automatically provide feedback to the construction site for real-time asphalt pavement construction quality control. The entire process works automatically without much human intervention and covers the entire project throughout the entire construction period in real time. The monitoring system was successfully implemented in the ZhaoMa Highway construction project in China, where it proved to be effective and reliable. The key asphalt plant production parameters, including asphalt content and aggregate gradation, were improved to a statistically significant level as shown by comparing before and after the monitoring system's deployment. Statistical analysis between asphalt core density and laydown and compaction monitoring data, collected at 20 locations, indicated that rolling passes were highly correlated with density, whereas finishing rolling temperature and HMA laydown temperature were moderately correlated with density. These results imply that implementation of the monitoring system proposed by this paper could improve asphalt construction procedure quality control and asphalt pavement construction quality.

Since the first highway was established in China in 1988, the construction of highways throughout the country has developed rapidly. According to the Chinese government's statistics, by 2015, total highway mileage had reached 120,000 km (*1*).

Unfortunately, highway asphalt pavements have suffered widespread premature failures. In addition to inadequate design and environmental issues, poor construction quality is one of the major causes contributing to premature failure. Since 2007, Chinese researchers have conducted extensive research work with regard to asphalt pavement quality. One area of these efforts is in the field of quality control of asphalt pavement construction.

Many factors are associated with pavement construction quality, with these grouped into two broad categories. External factors refer to the impact of raw materials, construction-related machinery, and climate. Technologies relate to the raw material production and asphalt paving construction equipment, the development of which has made satisfactory progress in China in the last two decades. The advancements in asphalt binder and aggregate production technology have significantly improved the quality of raw materials. The problems of asphalt pavement construction, such as mixture segregation, have been more effectively controlled through the use of large mixing plants, large pavers, and other advanced equipment.

The human-related factors are those that relate to the skills and proficiency of the engineers and workers involved in paving construction operations. It is not surprising that contractors may make some alterations during construction for profit purposes, such as changing production parameters in the mix production plant, altering the compaction procedure during the compaction operation, or even falsifying test data (*2*). Results of research studies have demonstrated that the abovementioned factors are the cause of a considerable portion

[1]MOE Key Laboratory for Urban Transportation Complex Systems Theory and Technology Beijing Jiaotong University, Haidian District, Beijing, China
[2]National Engineering Research Center of Road Maintenance Technologies, Haidian District, Beijing, P. R. China
[3]Research Institute of Highway, Ministry of Transportation, Haidian District, Beijing, P. R. China
[4]Beijing Zhongjiao Ruida Technology Co. Ltd., Beijing, China

Corresponding Author:
Address correspondence to Dongxiao Yu: sieistsonne@hotmail.com

of pavement construction quality problems. Therefore, efforts to control the human-related factors have become even more urgent to improve pavement quality control in China (*3*).

Although a well-established pavement construction quality control system exists in China, its monitoring procedures have two shortcomings: firstly, despite the use of some automatic detection equipment, such as the Pavement Quality Indicator for hot mix asphalt (HMA) density measurement in the field, the system is highly dependent on manual data collection, with this having low cost-efficiency, as well as being inaccurate. Secondly, manual quality control can test limited samples of the project rather than monitoring and testing the entire project. As the sampling rate is low, resolution of data is not as high as required by quality control.

To overcome these shortcomings, an asphalt pavement construction monitoring system has been developed and is presented in this paper. The system developed and presented in this paper can automatically collect, analyze, and deliver several key parameters in HMA production and field paving and compaction operations, automatically providing feedback of the pertinent online information to operators at the construction site. The system consists of two components: (1) HMA production monitoring and (2) field paving and compaction monitoring. In 2015, the system prototype was implemented and tested at ZhaoMa Highway construction project, Yunnan Province, China.

HMA Production Monitoring Component

HMA quality is one of the most important factors affecting asphalt pavement quality. It is also a critically important quality control point. With comprehensive standard specifications and corresponding inspection systems adopted nationwide for raw materials used for asphalt pavement construction in China, the quality check of aggregates, asphalt binders, and other additives used for producing HMA has matured. However, during HMA production in an asphalt plant, contractors may intentionally or unintentionally deviate from the proper production process, thus resulting in poor HMA quality. The standard HMA plant quality control processes, which include sampling and testing the asphalt mix once or twice a day (*4*), are not frequent enough to detect these problems. However, as the proposed system can simultaneously monitor certain key HMA plant operating parameters, discrepancies in HMA production can be detected with timely correction, to improve HMA production quality.

When devising an automated HMA production quality monitoring program, it is important to select only a few key parameters which critically influence HMA quality. Four key parameters were selected for the HMA plant monitoring system: asphalt content (asphalt–aggregate ratio); gradation; mixing temperature; and mixing time (*5*).

For monitoring aggregate gradation, it is neither necessary nor technically feasible to monitor all particle sizes ranging from 0.075 mm to 30 mm in the HMA produced in each batch. As the gradation of HMA produced in a batch plant (the predominant type used in China) can be accurately controlled by the amount of aggregates discharged from each hot aggregate bin, monitoring the discharge rates of aggregates from the hot bins has the same effect as monitoring the mix gradation itself. This assumption is based on the stability of aggregate gradations in the hot-bin gradations, which should be checked on a daily basis. As the HMA properties are more sensitive to variations in the amount of finer aggregates than of coarser aggregates, narrower fluctuation ranges for discharge rates must be set for the hot bins with finer aggregates (<3 mm) than for those with larger size aggregates (*6*).

Just as the weights of aggregates from the hot bins and the mineral filler discharged into the pugmill, as well as the amount of asphalt used in each mixing batch, are readily available from the asphalt plant operation control system, so too are the asphalt mixing temperature and the mixing time also available from the plant operation control. Therefore, the four parameters selected for monitoring the quality of HMA production can be obtained directly from the existing asphalt plant's operation control database. All that needs to be done is to acquire these data, perform real-time analysis, and issue warnings to appropriate persons when outliers are detected in any of these four key parameters.

The warning messages can be sent via SMS (Short Message Service). In the production plant monitoring system presented, the warnings are categorized into three different severity levels. A high-level warning (greater severity level) is sent to personnel at higher management level. The warning rules can be different for different projects. Table 1 presents the hierarchy of warnings used for the ZhaoMa Highway project. The monitoring program continually checks the quality data of every last successive 10-plate mix, and determines the warning level by the qualification rates and extreme values of quality indexes (Table 1).

HMA Mixture Monitoring Component

Paving and Compaction Monitoring Component

Two of the most critical operations on the job site are the laydown and the compaction operations. According to the Technical Specifications for Construction of Highway Asphalt Pavements (*7*), various laydown and

Table 1. Pavement Construction Quality Warning Notification Rules for ZhaoMa Highway Project

Warning index	Triggering condition	Warning level
Asphalt–aggregate ratio	80% ≤ % pass design value < 90%	1
	60% ≤ % pass design value < 80% or any single batch deviation > 0.3%	2
	% pass design value < 60% OR any single batch deviation > 0.3%	3
Gradation	80% ≤ % passing < 90% of any sieve size OR deviation of 0.075 mm > 1% in any single batch	1
	60% ≤ % passing < 80% of any sieve size OR deviation of 0.075 mm, 2.36 mm, or 4.75 mm > 2% in any single batch	2
	% passing of any sieve size < 60%	3
Mix temperature in pugmill	80% ≤ % passing < 90%	1
	60% ≤ % passing < 80% OR any single batch exceeds reject temperature limit	2
	% passing < 60% OR at least 3 batches exceed reject temperature limit	3
Mix time	80% ≤ % passing < 90%	1
	60% ≤ % passing < 80% OR any single batch < 35 seconds	2
	% passing < 60% OR at least 3 batches < 35 seconds	3

compaction controls are specifically configured for the construction of different types of asphalt mixes.

The mix temperature and paver speed, and environmental conditions, such as air temperature, wind speed, and other weather-related parameters, are important factors affecting the laydown operation. In addition, it is recommended that the paver operates continuously, at a constant speed if possible (*8*). Thus, in this paving monitoring system, the paving stop time is calculated for every single paving unit set at 10-meter (m) intervals, which can be obtained from the continuous Global Positioning System (GPS) data. An alarm light at the top of the paver will blink to alert and alarm the field technicians if any out-of-specification laydown-related index value is detected.

Regarding the site compaction operation, adequate compaction equipment, suitable rolling techniques, and an adequate mixture rolling temperature are essential to ensure effective compaction to achieve the desired compaction results. Some advanced types of roller compactor are equipped with special vibration-sensing devices that can measure the degree of HMA compaction. However, the authors' experience in China of using these intelligent rollers shows that extremely complex and diverse responses are received from the sensors for HMA made with different aggregate types, different mixture types, and with different compaction temperatures. Based on these experiences, in the case of improper calibration, core density estimation by this method has an error rate of at least 20%. Even with proper calibration, the error rate is still approximately 5% (*9*). Considering that the qualified compactness of the HMA layer is commonly 93–96%, a measured compactness value with ± 5% error is evidently unacceptable. Furthermore, in real construction environments, it is impossible to set the calibrations with the various combinations needed to incorporate the different types of roller and HMA and with ever-changing compaction temperatures. Therefore, in practical terms, this is not an effective way of monitoring compaction quality.

The rolling process to achieve the desired HMA compaction efforts involves two aspects: the rolling temperature and the rolling efforts generated by different types of rollers. The rolling efforts include using a particular set of rollers (static and vibratory steel-wheeled rollers and pneumatic-tired rollers), and following the rolling patterns established during the construction of the project's test sections. The rolling temperature requirements at each stage of the rolling operations are important and should be monitored. The need to control these parameters to achieve the desired compaction efforts is well understood and has always been an important part of process control in traditional quality control for asphalt paving operations. As good practice, the field quality control persons should ensure that this process is closely followed during paving operations. However, the problem is that this important process has rarely, if at all, been explicitly monitored on a real-time basis because of the lack of available techniques to do so. The proposed compaction monitoring system presented in this paper can provide this capability.

Development of the Monitoring System

Main Architecture. The major challenge today in automating the control process is the automated measurement of project performance indicators (PPIs). The effective control of PPIs needs two types of information in real time: (1) a list of the activities to be performed on the given day broken down in terms of PPIs and (2) measurement of the actual performance in the same terms. The first type of information is automatically extracted from the project model, as shown in Table 1, with up-to-date project planning and design standards. The proposed quality control system collects the latter type of information in real time. Moreover, the information is stored and analyzed with feedback provided to operators. Conducting such a process manually is complicated and labor intensive—it requires extensive monitoring. The proposed system design and its development process have focused on practicability. Consequently, this paper proposes a prototype of a quality check system which is stable, reliable, user-friendly, and reproducible (Figure 1).

The system functionality consists of three layers:

- Data collection layer: This is the basic layer, responsible for data collection and uploading in real time.
- Data analysis layer: This is the core component, formed by the database and the central logic program. This program is developed in web service mode, which does not directly interact with users, but provides a variety of function interfaces to the service layer. When a request from the service layer is received, the central logic program acquires relevant data from the database and completes all necessary computations, then submits the results back to the service layer. Thus, the main task of this layer is data processing and computing.
- Service layer: This layer, which directly communicates with users, is formed by the web pages and SMS message warning program. The web pages are responsible for requesting data upon the user's instruction, and showing the results using graphical user interface. The SMS warning program is responsible for constantly examining the latest data and sending SMS messages to certain people when necessary.

Under this architecture, the tasks of each module are clear, and integration of tasks with each other reduces the tasks overall. Internal optimization, maintenance, or improvement in any module will not affect the normal working of other modules or the entire system.

When deploying this monitoring system to a different construction project, the Data Collection Layer and the Service Layer will need to be modified, for instance, by installing devices on new equipment or redesigning the web pages and warning program so the system is adapted to the new working environment. However, no change will be needed for the Data Analysis Layer. Therefore, multiple projects can use the same database and the same central logic program. This greatly simplifies replication and makes the system easier to be used for other projects.

Hardware Modules for HMA Quality Control

HMA Production Plant Monitoring Device. The hardware for monitoring HMA production utilizes a standard industrial personal computer (IPC). The improvements made to the standard unit comprise installation of specially developed data collection software and a 3G wireless network card. This compact unit is resistant to vibration, dust, and humidity. It supports automatic restarts when unexpected power outages occur, and is particularly robust to the in situ construction environment.

The IPC is connected to the plant controlling computer from where it automatically reads production parameters in real time (Figure 2). Simultaneously, the IPC transmits the data directly to the remote database via 3G communication. Data transmission is usually available when the unit is within SMS signal coverage, which applies to the vast majority of China.

HMA Transportation Monitoring Module. The HMA transportation monitoring module is designed to fill the monitoring gap between HMA production and HMA paving. The most critical role of this module is to associate the HMA production batch number with the location where the batch of HMA is being used in paving and compaction. To fulfill this requirement, two sensors, namely, the GPS sensor and the infrared sensor, are mounted on HMA transportation dump truck. The GPS sensor for HMA transportation monitoring does not need high-precision positioning. It tracks the dump truck and records the timing of HMA loading and the loading duration. The infrared sensor is installed on the bottom of the dump truck tipper. It can detect the tilt movement of the dump truck. Therefore, the timing of the HMA unloading can be obtained. The transportation monitoring module has not been applied in the field in the studied highway project, because of the short distance from production to the construction site (approximate 10 km) and the contractor's truck fleet management.

HMA Paving and Compaction

Roller Positioning Module. The compaction level depends, among other factors, on the number of passes. Extensive work has been done in the current study to develop systems that monitor the trajectories of the compactor's movement and display them visually to the operator.

This model of the roller positioning module was realized in a prototype system and tested on a road construction site, using a GPS antenna mounted on each piece of equipment performing the controlled activities. The GPS assembly includes a fixed GPS base station and a portable GPS set attached to a roller. Using the GPS differential technology, the real-time trajectories of the roller can be measured to the decimeter level, with this meeting the positioning accuracy requirements for monitoring the rolling process (*10*).

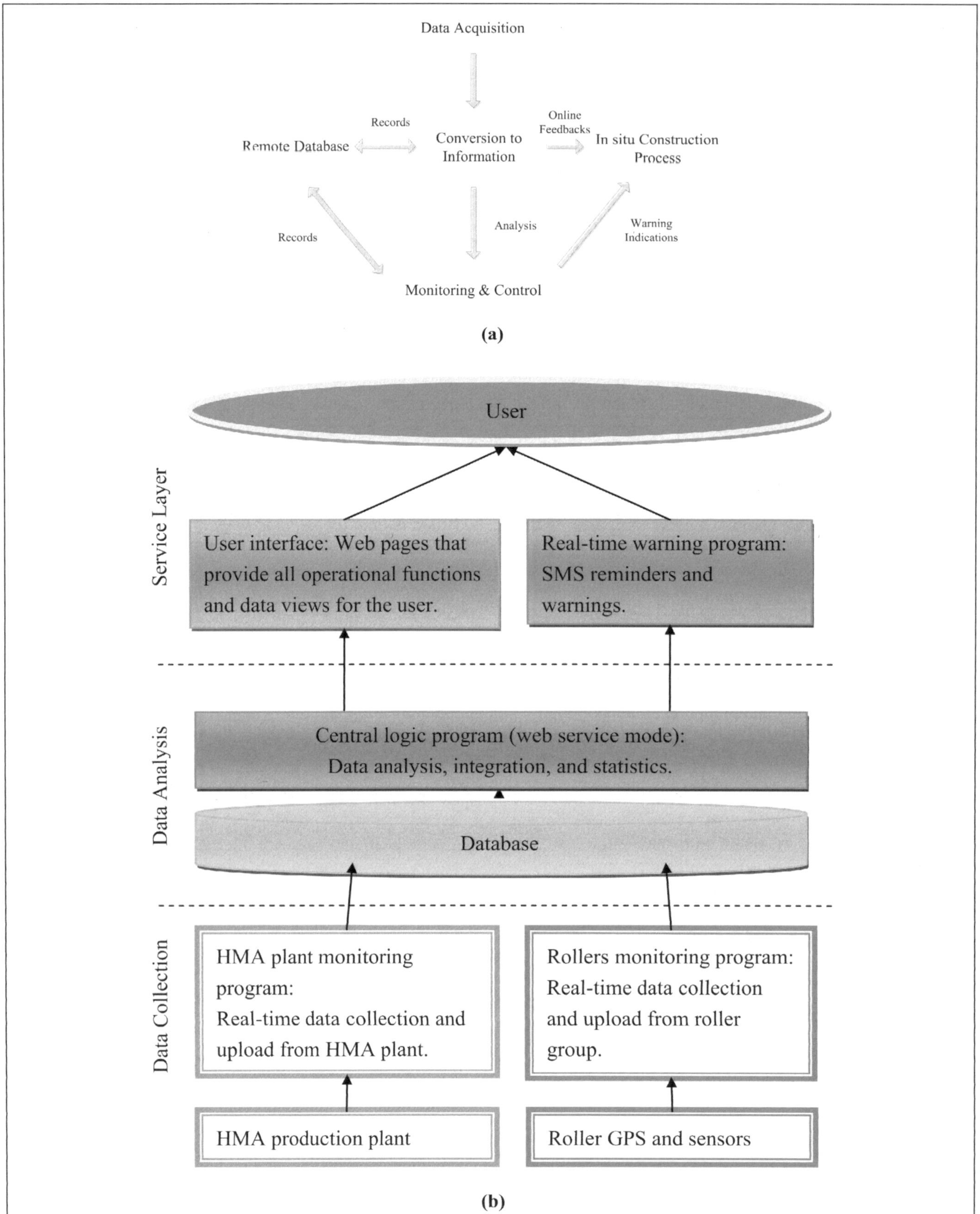

Figure 1. (*a*) Conceptual design of asphalt construction monitoring system; and (*b*) main architecture of pavement construction monitoring system.

Figure 2. Equipment on roller (*a*) controlling computer; and (*b*) detachable GPS antenna.

The GPS antenna is set on the top of the roller and can use its power supply. The effective radius of the GPS base station is approximately 10 km (Figure 2*b*). Therefore, the GPS base station will have to relocate for the paving project's next position which is over 20 km away.

Temperature and Vibration Sensors for Monitoring Paving and Compacting Operations. The temperatures of both the mix behind the paver's screed and the pavement surface are monitored by a non-contact sensor placed on the bottom side of the roller, approximately 0.5 m above the ground's surface, with this sensor able to measure the pavement surface temperature to within 1°C. To avoid biased temperature measurement, the sensors were calibrated. A simplified linear model is proposed which can estimate the internal pavement temperature from the air temperature and the measured pavement surface temperature (*11*). The model is

$$Y = ax + b \tag{1}$$

where

a = the coefficient;

b = the intercept;

x = the temperature measured by paver-mounted sensor; and

y = the HMA temperature measured by thermometer.

Vibration is measured by a vibration sensor mounted on the side of the roller's axle. The vibration-monitoring component is essentially an accelerometer. The three states of vibration, namely, heavy vibration, light vibration, and non-vibration are classified by the acceleration quantity collected by the accelerometer (Figure 3). The vibration mode was applied to all rollers working at both highway worksites of SuoMeng and ZhaoMa that were chosen for a before-and-after comparison.

Table 2. Flickering Indicators on Roller and Paver

	On roller	On paver
Green	Normal	Normal
Red	Low surface temperature	Low HMA mix temperature
Yellow	Speed too fast	Speed too fast
White	–	Unsuitable weather condition

Human–Machine Interface: Vehicle Screen and Alarm. As reported in research conducted by Beainy et al. (*12*), a vehicle screen mounted in the roller driving room displays the number of rolling passes and ambient temperature, which helps roller drivers to pay attention to the areas with an insufficient number of rolling passes or an abnormal temperature. Thus, the operator can appropriately adjust the rolling operation accordingly.

An alarm lamp is set on top of the roller and paver. When an unqualified index is detected, the lamp flickers with various colors to notify the field technicians (Table 2).

Pretreatment of GPS Monitoring Data

The hardware system described above can collect the raw data. However, some raw data need to be processed to yield meaningful results. The most important ones are the GPS coordinates that need to undergo transformation and the computation of various indicators. A schematic of the entire HMA pavement and compaction quality inspection system is shown in Figure 4.

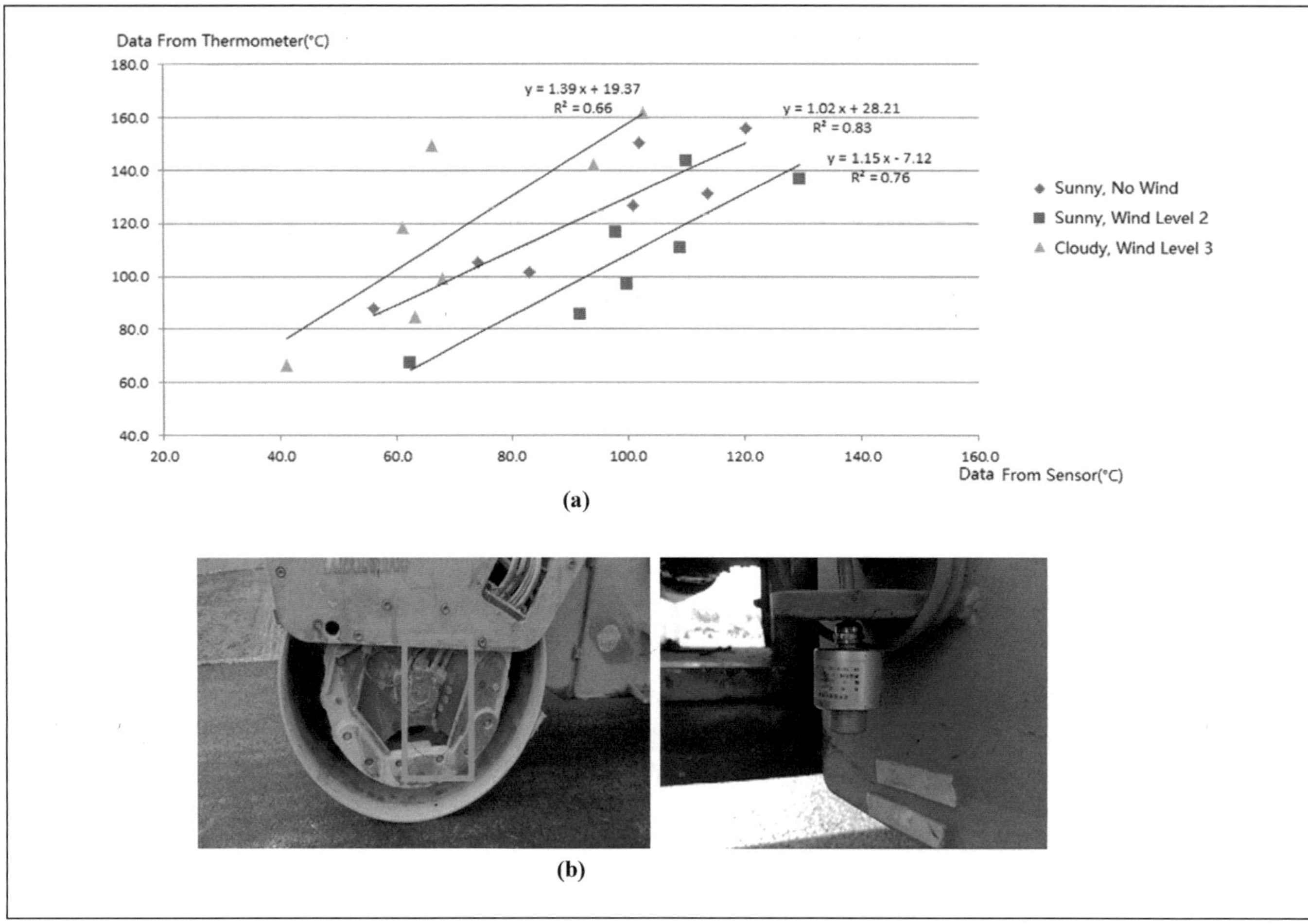

Figure 3. (*a*) Asphalt pavement temperature sensor; and (*b*) vibration sensor on axle of roller.

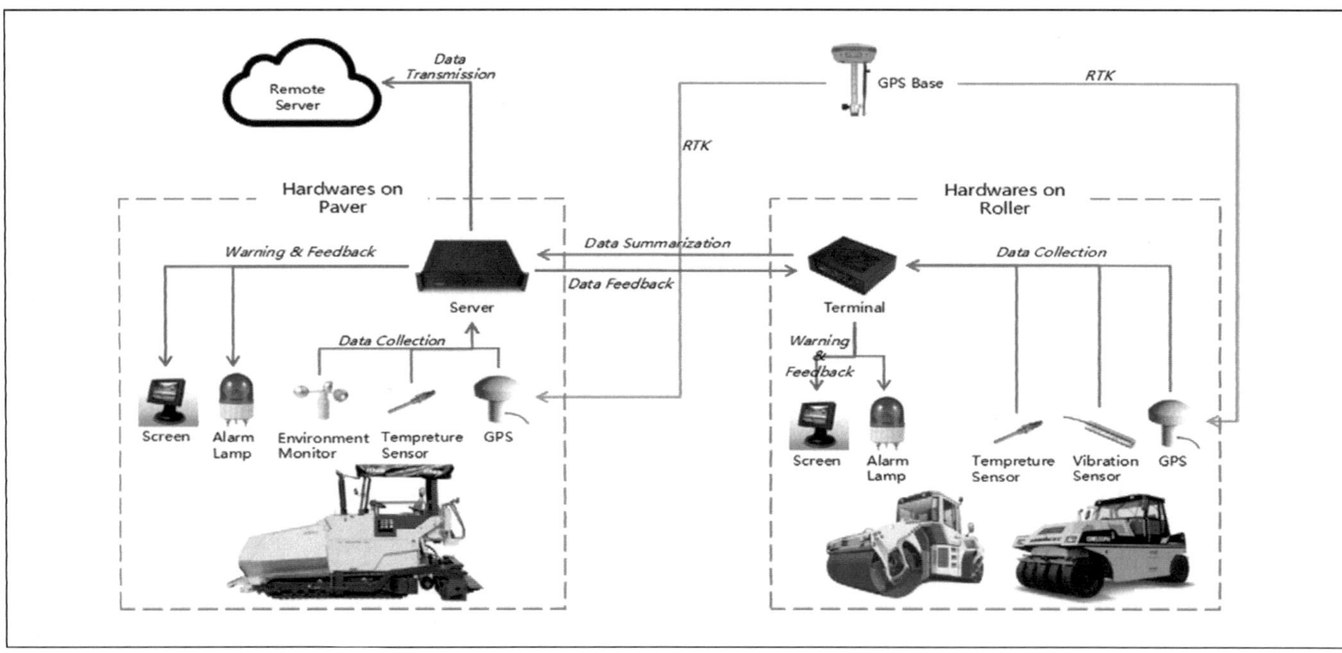

Figure 4. Composition of HMA paving and compaction quality check system.
Note: RTK = real-time kinematic.

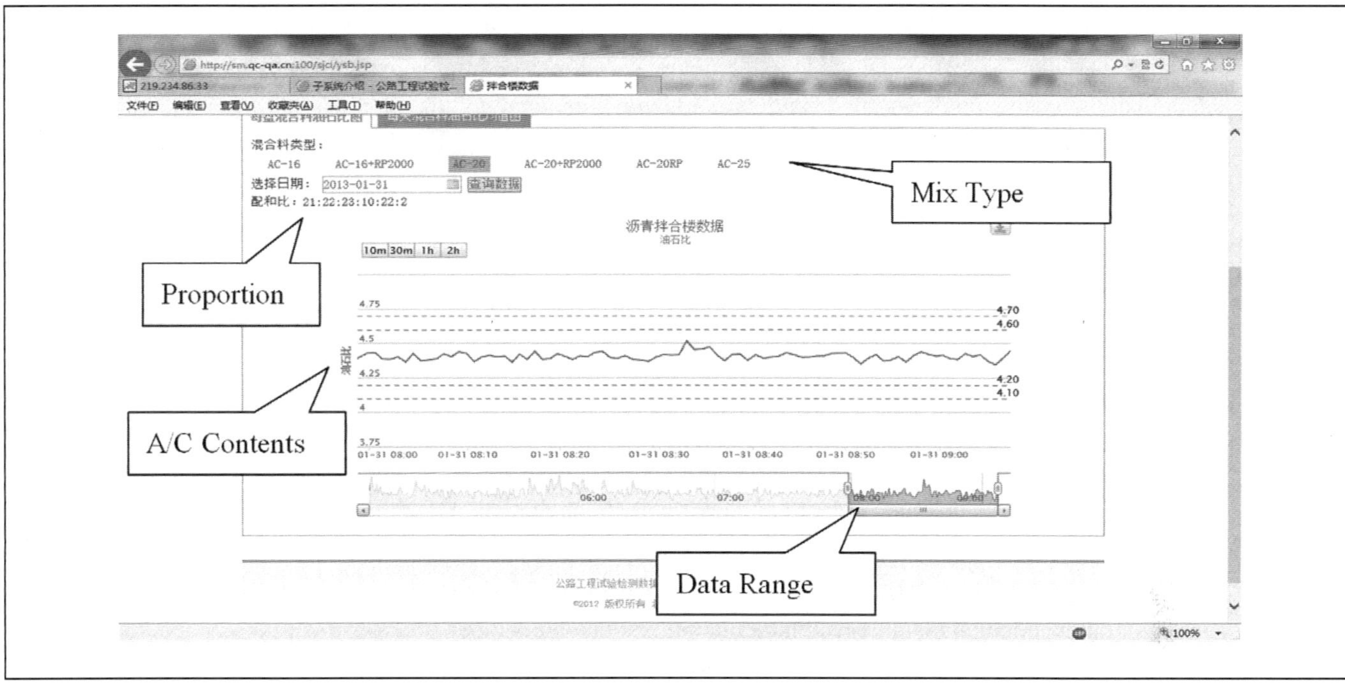

Figure 5. Graphical data view of asphalt–aggregate ratio.
Note: A/C Contents = asphalt content.

GPS Coordinates' Transformation. The GPS antennae mounted on both the paver and the roller provide the latitude and longitude coordinates of the equipment location. Therefore, the latitude–longitude coordinates obtained from the GPS antennae should be transformed into the pavement coordinates represented by the milepost number and cross distance (the transverse distance from roller to highway centerline). In general, construction design documents include a corresponding table of milepost numbers and latitude–longitude coordinates of the road centerline. Therefore, the coordinates' transformation can be finalized by tabulation, provided the design documents and the GPS antennae on the paver and roller both use the same coordinate system.

Data Alignment. In order to provide useful information for quality control to work more efficiently, the monitoring data of the paver are arranged in units of every 10 meters, including the average paving speed, average mix temperature, stop counts, and stop time in each 10-meter unit (*13*).

The monitoring data on the roller are realigned, and information on the pavement is displayed on the screen by cells, each one representing a 1 m × 1 m pavement area. When the GPS antenna collects the data from each roller and sends these data to the server, they are overlaid in real time and the number of rolling passes of each 1 m × 1 m cell is calculated instantaneously.

Software Platform

Functions of HMA Plant Monitoring. The back-end software is developed based on a browser/server (B/S) structure, in which all applications and functions are operated on web pages. Users have different privileges with different functions on the web pages. The detailed functional features include:

- Automatic update data every five minutes to browse the latest detailed data transmitted from the HMA production plant;
- Online acquisition of statistics, such as total amount of production and the passing rate;
- Graphical display monitoring results of key parameters, such as the data view of asphalt–aggregate ratio from each plant mix (Figure 5).

Functions of Roll Compaction Monitoring. The compaction monitoring system dynamically presents the data of the various roller operation-related indicators (Figure 6*a*). A table at the top of the web page shows roller- and operation-related parameters, which are updated in real time during the compaction operation. The figure at the bottom of the web page presents the rolling track on the pavement. The black lines symbolize the pavement edges. The blue area is the rolling track which also has updates in real time. The number of rolling passes on the paving is presented through the depth of color. A deeper color

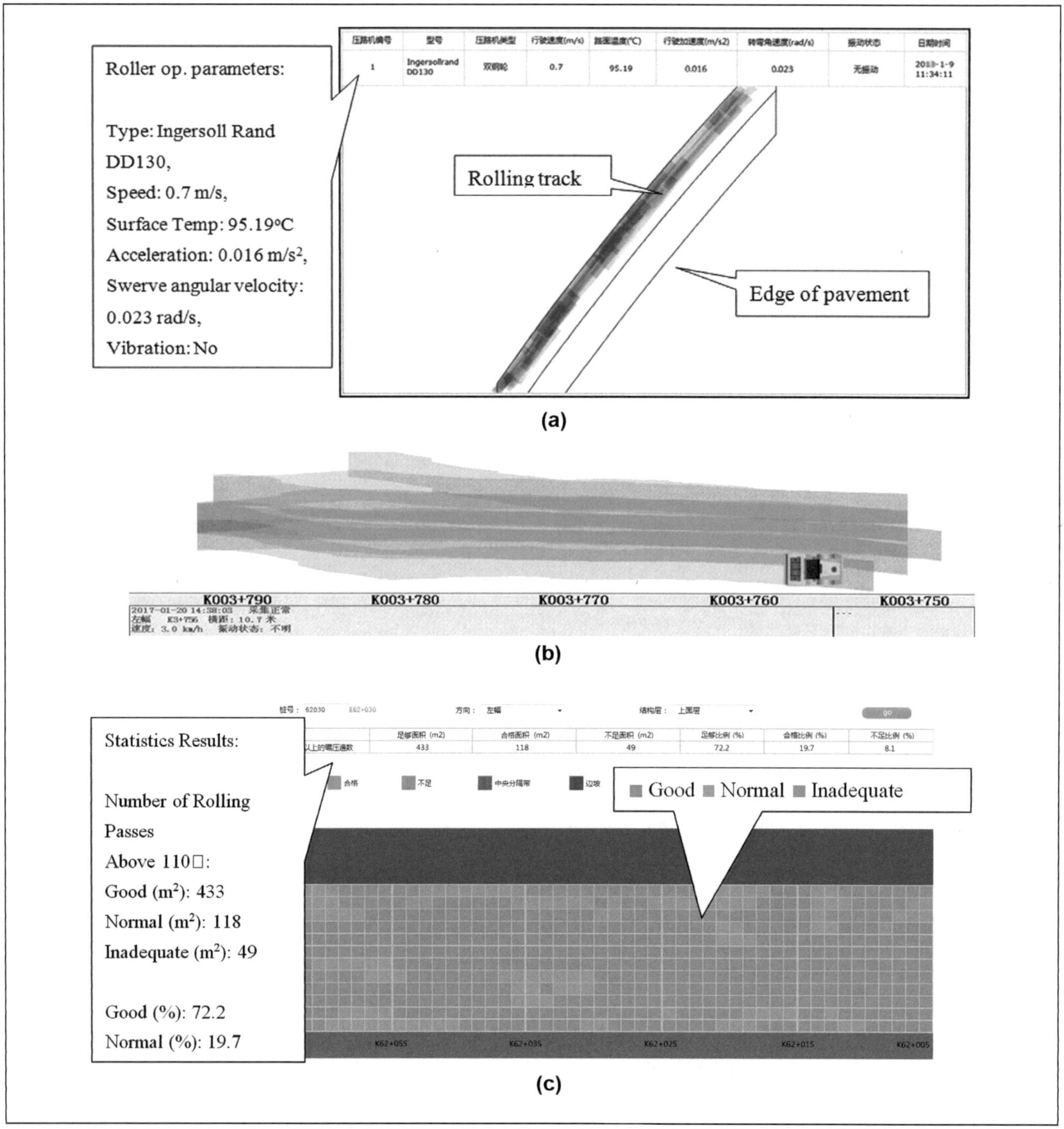

Figure 6. (*a*) Web-based roller real-time working status; (*b*) roller driver working status user graphic interface; and (*c*) graphic display of rolling quality statistics.

means a higher number of rolling passes (Figure 6*b*). Thus, the operator can cover the entire pavement area and the exact number of rolling passes, thereby preventing anything from being missed.

Comprehensive Analysis. The asphalt pavement construction monitoring data are stored permanently in the system's database after the construction is completed. This dataset can be queried and analyzed online. For example, the number of rolling passes, the pavement surface temperatures, and the rollers' vibration state at any position within the construction site can be queried and analyzed to obtain the corresponding pavement compaction quality at those positions: the detailed statistical results can be presented in 1 m × 1 m cells (Figure 6c). These data are valuable for subsequent pavement maintenance management after the pavement has been opened to traffic (*14*).

Before-and-After Comparative Analysis

Three indicators, namely, mixing temperature, HMA gradation, and density, were selected to test the performance of the quality control system. The aim was to test whether it was possible to improve HMA paving and compaction operations through a quality check system that monitors and sends feedback to operators at the construction site. The asphalt pavement construction on the ZhaoMa Highway started in May 2015, and was completed in November 2015. This monitoring system was deployed in this project and began operating in late August 2015. Data from the key parameters in the asphalt plant operation were collected before and after the monitoring system was deployed and were compared to assess whether any significant effects could be observed (Table 3).

For pavement construction, core density is a critical quality indicator. In total, 20 core samples were collected from random positions on a segment of ZhaoMa Highway's asphalt surface layer (AC13 with SBS [styrene-butadiene-styrene] asphalt, 4 cm depth). The core density measured for these 20 samples and the corresponding laydown and compaction monitoring data collected by the quality control system at these 20 locations are presented in the Appendix. On the other hand, samples had also been collected at another construction site located in the SuoMeng Highway project, Yunnan province, where the contractor, also in charge of the ZhaoMa Highway project, was not using the proposed quality check system.

Therefore, to investigate the significance of the means of core density, the independent two-sample *t*-test was applied (*15*). Specifically, the *t*-test was used to determine whether core densities, collected from two pavement construction sites, differed from each other at a statistically significant level after the application of the quality check system. The basic assumptions taken into account were: (a) the samples are unpaired, and (b) the sample differences can be viewed as a random sample from a population of differences. The results of the independent *t*-test are summarized below:

Table 3. Before-and-After Comparison of Pavement Construction Quality Data

Indicators	SuoMeng (Before) Sample size	ZhaoMa (After) Sample size	Statistical difference (5%)
Core density	20	69	Significant
Gradation	34	66	Significant
Mixing temperature	80	122	Insignificant

Null hypothesis: H0 = hypothesized value = 0

$$\mathrm{t} = \frac{\boldsymbol{m}_A - \boldsymbol{m}_B}{\sqrt{\frac{S^2}{n_A} + \frac{S^2}{n_B}}} \tag{2}$$

$$S^2 = \frac{\sum (x - \boldsymbol{m}_A)^2 - \sum (x - \boldsymbol{m}_B)^2}{n_A + n_B - 2} \tag{3}$$

$$\mathrm{df} = n_A + n_B - 2 \tag{4}$$

where

A and B: the two groups being compared;

m_A and m_B: the means of groups A and B, respectively; and

n_A and n_B: the sizes of groups A and B, respectively.

Likewise, statistical analysis using the independent two-sample *t*-test was carried out on mixing temperature and HMA gradation (Table 3).

Two of the three key asphalt plant production parameters, comprising asphalt content and aggregate gradation, were significantly improved after the monitoring system was deployed. On the other hand, the mix production temperatures were close to uniform, passing the target mix temperature both before and after the plant monitoring system was deployed. The results of these improvements indicate that the system delivers effective information to plant operators and quality control personnel on whether standard operation processes are being followed. It should be mentioned that the contractor was informed once the monitoring system was deployed.

Conclusion and Future Work

The asphalt pavement construction monitoring system presented in this paper covers three main asphalt pavement construction processes: HMA plant production, HMA transportation, and HMA laydown and compaction operations. The main architecture of the monitoring system, the development of the required hardware and the software structure, and the implementation process of the monitoring system are presented in this paper.

The system is able to perform automatic data collection, data transmission, and quality risk identification, and can automatically provide feedback to the construction site for real-time asphalt pavement construction quality control. Alongside the HMA transportation module, the other modules work automatically, monitoring the pavement construction activities throughout the entire construction period.

The monitoring system was successfully implemented in the ZhaoMa Highway project, China, where it was proven to be cost-effective and reliable. A comparison of the HMA production quality before and after implementation of the plant monitoring system showed that key asphalt plant production parameters, comprising asphalt content and aggregate gradation, were significantly improved after the monitoring system was deployed. Statistical analysis of asphalt core density collected from varied road segments indicated that the system had significant positive effects on construction quality. These results imply that implementation of the monitoring system proposed by this paper could have the potential to improve asphalt construction monitoring capability, thus improving asphalt pavement construction quality. However, pavement construction involves several factors, some of which are not under system monitoring.

The main contributions of the proposed monitoring system platform is its ability to carry out detection in real time and to issue warnings of certain abnormalities that are occurring during asphalt pavement construction, thus allowing timely corrections. More importantly, the fact that the entire set of pavement construction monitoring data can be stored permanently and can be queried and analyzed at any time thereafter has important positive implications for pavement maintenance and management during the operation period. Under China's current pavement management practice, an entire pavement project is viewed as homogeneous in quality. This means that the inherent variability in pavement quality has often been neglected; thus, the same maintenance and rehabilitation practices are applied to the entire project. The availability of the complete set of pavement construction monitoring data for the project and the incorporation of this information into the pavement management decision process would be very beneficial in developing more targeted maintenance and rehabilitation strategies for different parts of the project, which would be much more objective and effective. This constitutes the authors' next objective: to develop an integrated pavement construction–maintenance lifecycle management system.

Appendix Sampling Locations

ID	Sampling location
1	Right K042+742
2	Left K071+054
3	Left K075+585
4	Right K057+877
5	Left K060+429
6	Left K070+383
7	Left K073+754
8	Left K079+534
9	left K072+734
10	Left K049+509
11	Left K047+962
12	Right K077+350
13	Right K070+549
14	Right K064+792
15	Right K075+782
16	Left K052+841
17	Left K048+643
18	Left K078+334
19	Left K065+189
20	Left K079+345

Acknowledgments

This research was supported by fundamental research funds for the central universities, No T15RC00050.

Author Contributions

The authors confirm contribution to the paper as follows: study conception and design: Sicong Zhu, Dongxiao Yu; data collection: Xiangdi Li; analysis and interpretation of results: Haoyang Wang; draft manuscript preparation: Sicong Zhu. All authors reviewed the results and approved the final version of the manuscript.

References

1. Ministry of Transport of the People's Republic of China. *Technical Specification for Construction of Highway Asphalt Pavement*. China Communications Press, Beijing, China, 2004.
2. Xiuwen, M., G. Li, and Z. Jinsheng. Discussion on Quality Control Standard of Asphalt Pavement Compaction Degree. *Forest Engineering*, Vol. 6, 2011, p. 017.
3. Tan, Y., L. Yao, H. Wang, Q. Du, and Q. Dang. Analysis of Asphalt Pavement Structural Premature Damage Position and Countermeasure. *Journal of Highway and Transportation Research and Development*, Vol. 5, 2012, p. 002.
4. Aschenbrener, T., B. Schiebel, and R. West. *Three-Year Evaluation of the Colorado Department of Transportation's Warm Mix Asphalt Experimental Feature on I-70 at Silverthorne, Colorado*. NCAT Report 11. National Center for Asphalt Technology, Auburn, AL, 2011.

5. Hicks, R. G., and J. A. Epps. Quality Control for Asphalt Rubber Binders and Mixes. Rubber Pavements Association, Tempe, AZ, 2000.
6. Iskender, E., and A. Aksoy. Field and Laboratory Performance Comparison for Asphalt Mixtures with Different Moisture Conditioning Systems. *Construction and Building Materials*, Vol. 27, No. 1, 2012, pp. 45–53.
7. Highway China. *The 2015 Chinese National Highway Mileage List*. http://www.china-highway.com/News/bencandy/id/112770.html. Accessed December 28, 2015.
8. Ter Huerne, H., A. Dorée, and S. Miller. Monitoring Hot Mix Asphalt Temperature to Improve Homogeneity and Pavement Quality. *Proc., 6th International Conference on Maintenance and Rehabilitation of Pavements and Technological Control (MAIREPAV6)*, Torino, Italy, 2009, pp. 556–565.
9. Al-Qadi, I., Z. Leng, S. Lahouar, and J. Baek. In-Place Hot-Mix Asphalt Density Estimation Using Ground-Penetrating Radar. *Transportation Research Record: Journal of the Transportation Research Board*, 2012. 2152: 19–27.
10. Peyret, F., D. Betaille, and G. Hintzy. High-Precision Application of GPS in the Field of Real-Time Equipment Positioning. *Automation in Construction*, Vol. 9, No. 3, 2000, pp. 299–314.
11. Diefenderfer, B. K., I. L. Al-Qadi, and S. D. Diefenderfer. Model to Predict Pavement Temperature Profile: Development and Validation. *Journal of Transportation Engineering*, Vol. 132, No. 2, 2006, pp. 162–167.
12. Beainy, F., S. Commuri, and M. Zaman. Quality Assurance of Hot Mix Asphalt Pavements Using the Intelligent Asphalt Compaction Analyzer. *Journal of Construction Engineering and Management*, Vol. 138, No. 2, 2011, pp. 178–187.
13. Miller, S., T. Hartmann, and A. Dorée. Measuring and Visualizing Hot Mix Asphalt Concrete Paving Operations. *Automation in Construction*, Vol. 20, No. 4, 2011, pp. 474–481.
14. Zimmerman, K. A., L. M. Pierce, E. Perry, and N. Saadatmand. The Development of a Pavement Management Roadmap. Presented at 90th Annual Meeting of the Transportation Research Board, Washington, D.C., 2011.
15. Peck, R., and J. L. Devore. *Statistics: The Exploration & Analysis of Data*. Cengage Learning, Boston, MA, 2011.

The Standing Committee on Emerging Design and Construction Technologies (AFH30) peer-reviewed this paper (18-04397).

Article

TRR
JOURNAL OF THE TRANSPORTATION RESEARCH BOARD

Transportation Research Record
2018, Vol. 2672(26) 40–47

DOI: 10.1177/0361198118755665
journals.sagepub.com/home/trr
SAGE

First Application of UHPC Bridge Deck Overlay in North America

Sri Sritharan[1], Gaston Doiron[2], Dean Bierwagen[3], Brian Keierleber[4], and Ahmad Abu-Hawash[3]

Abstract

Deterioration of existing bridge decks, which usually originates with the deck cracking on the top surface, is a common problem in North America. It causes frequent repair of the decks to limit further damage resulting from water/chloride ingress. Superior engineering and durability properties facilitate the use of ultra-high performance concrete (UHPC) as an attractive alternative for a deck overlay, minimizing both deck deterioration and maintenance costs. Recently developed UHPC thixotropic mix designs, which are different from commonly used self-leveling UHPC, enable UHPC overlay to be used on decks with slopes and meet specific crowning requirements. The use of a UHPC thixotropic mix design with 3.25% of steel fibers was successfully evaluated under laboratory conditions by applying it on sloping deck surfaces with appropriate roughness between the normal concrete (NC) and UHPC. The feasibility of applying this technology in the field was then investigated on a small bridge for the first time in North America in May 2016. This paper presents the details about the laboratory evaluation, field implementation of UHPC overlay, and lessons learned from this first UHPC overlay project in North America.

Bridge deck deterioration is a frequent problem faced by many states in the U.S., especially the northern states where deicing is used and coastal states. Maintenance and repair associated with bridge decks can cost 50–80% of all bridge-related expenditures, which is conservatively estimated to be more than US$5 billion per year (*1*). Typical bridge deck deteriorations include cracking, spalling, delamination, and corrosion of reinforcement (e.g., Figure 1). Various phenomena contribute to deck deterioration, including poor initial quality, the use of deicing salts in winter, overloading, stresses associated with freeze-thaw cycles, fatigue, corrosion of reinforcement, or any combination thereof. Near the coast, the bridge decks can experience damage due to seawater salt. Several of these problems initiate due to surface cracking and infiltration of moisture and chloride into the deck.

A variety of techniques are used to repair damaged bridge decks. They can range from grouting to seal surface cracks to partial/full depth patching to replacing partial/full bridge deck. While each of these techniques can increase the service life of the bridge decks, none has been proven to completely prevent further bridge deck deterioration, nor do they increase the longevity of bridge decks.

With its superior durability properties, the use of ultra-high performance concrete with fiber reinforcement (UHPC or UHPFRC) in bridge decks has been explored in Iowa as a means to combat wide-ranging bridge deck deterioration problems. Given that solid UHPC slab deck is prohibitively expensive, use of a waffle deck was successfully demonstrated (*2, 3*). By optimizing the waffle deck configuration, it was shown that the waffle deck concept can be used to reduce the volume of the required UHPC in bridge decks by 50% or more in comparison to solid bridge deck panels (*4*). To further reduce the construction cost and broaden the application of UHPC in bridge decks, use of a thin UHPC layer as an overlay was then explored with an intention of preventing deck cracking, reducing fatigue damage and minimizing penetration of moisture and chloride ions into the bridge deck. To ensure cost effectiveness of this concept, the thickness of the UHPC layer was targeted to be about 1.5 in. with no mechanical connection between the UHPC layer and the underlying normal concrete (NC) deck. This approach will reduce the required volume of UHPC in bridge decks to less than 20% when compared with a solid UHPC panel of nominal thickness. A systematic investigation involving UHPC-NC slant shear tests and flexure and shear tests on a 9 ft (long) × 2 ft (wide) concrete deck with consideration to durability properties concluded the following (*5*):

[1]Department of Civil, Construction, and Environmental Engineering, Iowa State University, Ames, IA
[2]LafargeHolcim, Mississauga, Ontario, Canada
[3]Office of Bridges and Structures, Iowa Department of Transportation, Ames, IA
[4]Buchanan County Secondary Roads Department, Independence, IA

Corresponding Author:
Address correspondence to Sri Sritharan: sri@iastate.edu

Figure 1. Examples of damage to bridge deck (*1*).

Figure 2. Chillon Viaducts and placing of UHPC on bridge deck by WALO.

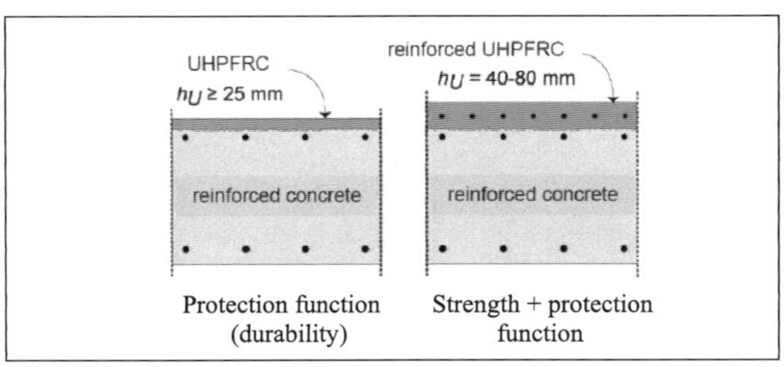

Figure 3. SIA 2052 recommendations for UHPC overlay (7; 1 mm = 0.0394 in.)

- A thin UHPC layer as an overlay has the potential to increase longevity of a bridge deck;
- To ensure composite action between UHPC and NC, a minimum interface roughness of about 1/8 in. is required;
- A modified UHPC mix is needed to ensure that UHPC can be placed on sloped surfaces and allow necessary crowning to be done in the field; and
- A long-term field evaluation of UHPC overlay is necessary to ensure the top surface or the UHPC-NC interface is not affected by freeze-thaw cycles and fatigue loading.

Focused on increasing both the strength and durability, UHPC-NC composites were explored in Switzerland (*6*). To enhance the strength of the composite deck, mild steel reinforcement is placed within the UHPC layer and this reinforcement mat is intended to be tied to the NC deck using vertical bars penetrating into the existing deck. This research led to field applications and development of a standard (*7*). The largest application of this effort was the rehabilitation of bridge decks of the Chillon Viaducts near Lausanne, Switzerland (see Figure 2). Completed in 1969, the viaducts consist of two 1.3-mile long concrete box-girder bridges and required refurbishment of a 62,191-square-yard surface area. This project was completed in 2015 by WALO, who developed special placing and batching equipment that could sustain the large volumes of UHPC (capacity of 11.8 cubic yards per hour) required for the job. LarfargeHolcim provided a UHPC mix with 3.25% steel fiber content for this project, in which an overlay thickness ranging from 1.6 in. to 2.0 in. was used.

As a guide for the design and specification of UHPC overlays, SIA 2052 (*7*) has recommended the appropriate mix based on the type of application. For UHPC Class UA or UB, the possible application of overlay is depicted in Figure 3. These two UHPC classes require strain-hardening characteristics such that they can withstand the development of eigenstresses due to restrained shrinkage without experiencing any cracking, and

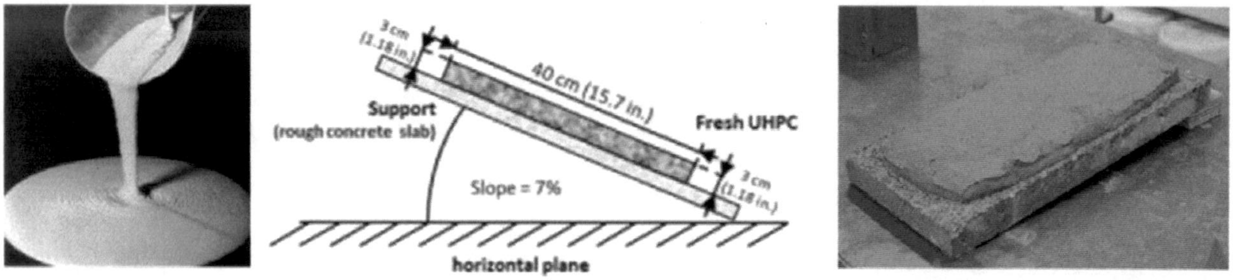

a) Flow test b) Slope test dimensions c) Slope test of a UHPC mix

Figure 4. Testing of UHPC rheology (1 cm = 0.394 in.)

Table 1. Basic Chararteristic Properties of the UHPC Overlay Mix (9)

Property	Typical value
Uniaxial tensile behavior: type	UA per MCS-EPFL (7)
Total shrinkage at 90 days	500 μstrain
Elastic tensile strength at 28 days	1.2 ksi
Tensile strength at 28 days	1.3 ksi
Strain when the tensile strength is reached (hardening)	0.35 %
Compressive strength on cube at 28 days	18 ksi
Modulus of elasticity (28 days)	6530 ksi
Water porosity at 90 days	6%
Diffusion coefficient of chloride ions at 90 days	$\leq 1.1\times10^{-12}$ $ft^2.s^{-1}$
Apparent gas permeability at 90 days	$\leq 5.3\times10^{-18}$ ft^2

*Commercially available version of the product is Ductal® NaG3 TX.

have a minimum bond stress of 0.22 ksi and a rheology compatible with the ability to place UHPC at a slope of up to 10%.

Establishing an Overlay Mix

UHPC Mix Consistency

A beneficial characteristic of UHPC is its ability to flow and self-level, which has been taken advantage of in previous projects completed in Iowa (*8*). Using the same type of consistency on a bridge deck would be challenging because most decks have slopes and cross-slopes. Therefore, LafargeHolcim worked on a new UHPC overlay formulation that can maintain a slope of up to 7% when placed on bridge decks. This formulation, which was comparable to the mix deployed on the Chillon Viaducts, used components available in the U.S. and included 3.25% of steel fibers. These fibers were 0.008 in. in diameter and 0.5 in. in length.

Standard slump or flow tests are not useful for a UHPC overlay mix because of its stiff rheology. A method to verify that the mix can hold a required slope is suggested in SIA 2052. By using a platform to replicate the slope of the bridge deck under consideration, the UHPC formulation can be adjusted for that specific slope (see Figure 4). The goal here is to use a formulation that is as flowable as possible to facilitate placing, while the mix also needs to hold the specified slope. By testing the material after batching, if it is found that the consistency requires an adjustment, this can be done before the material is used on the deck.

Strain Hardening

Critical characteristics of the overlay mix include impermeable surface and strain hardening, which typically require steel fibers. Steel fibers are costly and using them efficiently in a mix is important. Therefore, the amount of steel fibers used in the mix needs to be sufficient to allow stresses to be taken by the steel fibers without cracking the mineral matrix. Testing conducted by LarfargeHolcim suggested that 3.25% of steel fibers per volume would be optimal for the overlay application. The basic characteristics of this UHPC overlay mix as produced by LafargeHolcim are summarized in Table 1.

Laboratory Trial

To verify the quality of the UHPC overlay mix and provide an opportunity for a contractor to experiment with the mix, two 8 × 8 ft^2 square slabs were cast with typical deck reinforcement details. They both were 7.75 in. thick; the top surface of one of

Figure 5. UHPC overlay trial in the laboratory: (*a*) surface with exposed aggregates; (*b*) broom-finished surface; (*c*) placing of UHPC on one of the sloped slabs.

Figure 6. Mud Creek Bridge in Buchannan County, Iowa: (*a*) aerial view (*courtesy of Google*), and (*b*) condition of bridge deck prior to rehabilitation.

the slabs was finished with exposed aggregates and the other was constructed with a broom-finished surface (Figure 5, *a* and *b*). Both slabs were positioned at 6° of slope and UHPC overlay mix was batched and placed on the slabs in the structural engineering laboratory at Iowa State University (Figure 5*c*). This exercise confirmed that the selected UHPC overlay mix is satisfactory in terms of ensuring workability and holding the slope when placed on a roughened surface.

Prototype Bridge

After evaluating several candidate structures, the three-span, two-lane, Mud Creek Bridge located on Buchanan County Road D48 near Brandon, Iowa, was selected to trial the first UHPC overlay on a bridge deck in North America. This 102-ft long and 30-ft wide straight, continuous concrete slab bridge has a 5% superelevation and was built in the mid 1960s. An aerial view and the condition of the bridge deck surface prior to applying the UHPC overlay are shown in Figure 6.

Field Application

The UHPC overlay was applied to the deck surface of the Mud Creek Bridge in May 2016. A key driver for bridge overlay applications is the duration of the intervention. Traffic needs to be returned to the bridge in the shortest time possible. In this particular case, the targeted duration of two weeks was accomplished from closure to opening of the bridge for public use.

As the first step following the closure of the bridge, the top 0.25 in. of the deck surface was removed. To ensure a satisfactory surface roughness, the deck was then grooved along the bridge length, leaving a surface roughness with an amplitude ranging from 1/12 in. to 1/8 in. (i.e., 2 mm to 3 mm). Finally, the deck was thoroughly cleaned and adequately saturated with water before placing the UHPC overlay.

On Site Batching and Placing

For bridge projects in the U. S., batching of UHPC is typically performed on site by the contractor. Normally, small

Figure 7. Mud Creek Bridge deck: (*a*) deck after preparing the surface; (*b*) mock-up; (*c*) wire mesh in the negative moment region; (*d*) placing of the UHPC overlay.

Figure 8. UHPC overlay on Mud Creek Bridge deck: (*a*) grooving of the surface; (*b*) finished surface.

amounts of material are required and placing the material takes time, and hence batching output is normally not the main driving factor. For large overlay jobs, having a portable batching plant on site as demonstrated for the Chillon Viaducts project would be beneficial. However, for small to medium span bridges such as the Mud Creek Bridge, using large size portable mixers is considered sufficient. For this first overlay project, a pair of high shear pan mixers were used. Each mixer had the capacity to take a bulk bag of UHPC premix, yielding approximately 0.65 yd^3 of material per batch. The typical duration of loading and batching of the UHPC took approximately 20 minutes. The UHPC overlay mixing was done at one end of the bridge and transported to the working area using a mini concrete dumper.

To place the UHPC efficiently, the deck was divided into two regions along the longitudinal axis of the bridge and the overlay was placed separately for each lane on two different days without using any special detailing across the construction joint. After the UHPC overlay mix was spread on the surface, a vibratory truss screed was used to compact the UHPC and maintain an overlay thickness of 1.5 in. A local contractor, who executed the project, had no prior experience with UHPC, and thus experimented with placing of the UHPC overlay material on a flat ground surface prior to working on the deck. Although placing of the UHPC overlay on the first lane took longer, the contractor successfully placed the material on the second lane within 3 to 4 hours. This included some wait time for the batching of the UHPC. A series of pictures from the construction site are shown in Figures 7 and 8.

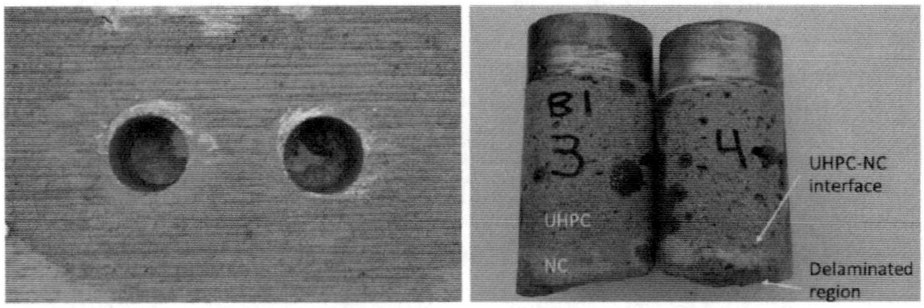

Figure 9. Samples of pull-off test conducted at the site by FHWA.

Curing

Hardening of UHPC normally takes a fair amount of time and is highly dependent on ambient temperature conditions. Using accelerators and heating are two alternatives that can help speed up the process. However, neither of these options are ideal for an overlay application. Heating is not practical for large surfaces, and using an accelerator would make batching and placing challenging. Therefore, neither options were used for the bridge deck overlay project. In North America, when UHPC has been used for joint closure pours applications, traffic is usually allowed back on the structure as soon as the UHPC gains a strength of 11.6 ksi. A similar approach was adopted for the UHPC overlay project. The compressive strength of the overlay placed on the second lane was 12.3 ksi at the age of 4 days, by which time the overlay placed on the first lane was 9 days old. Therefore, grinding and grooving of the UHPC deck surface was permitted after 4 days from placing the UHPC ovelay on the second lane.

Field Monitoring

The bridge deck has been subjected to periodic visual evaluations since the UHPC overdeck was installed. To identify potential delamination between the UHPC-NC interface, the deck was also subjected to chain dragging, which identified some areas for concern. Researchers from Federal Highway Administration (FHWA) conducted pull-off tests on areas identified to have both no delamination concerns and potentail delamination. The pull-off test is designed to quantify the strength between the repair material (i.e., UHPC) and the substrate material (i.e., NC) over a circular region with a 2 in. diameter. In areas where no delamination was suspected, the tests revealed no concerns for delamination since the tests did not cause any damage to the UHPC-NC interface. The failure occurred at the top of the UHPC where a metallic test disc was attched as part of the test procedure. In areas with potential concern, it was found that the delamination was within the old concrete deck layer beneath the interface and not at the UHPC and NC interface (see Figure 9). Other non-destructive evalautions of the bridge deck have been undertaken using thermal imaging technique. No concerns for delamination have been found. This effort will continue and findings will be reported in the future.

Lessons Learned

From concept development through installation, all parties involved in this first UHPC overlay application in North America put forward their best efforts and the result was a very successful project. Once a suitable bridge was identified, securing funding and coordinating with local contractors had posed some challenges because of the unfamiliarity with UHPC. Batching of the material on site using two portable mixers was satisfactory. However, the process could be optimized. A trained, sufficiently large crew to work with the mixers is important. Speeding up the batching by any means, for example by introducing the steel fibers more efficiently in the mix, would be helpful. For larger projects, the use of a portable batching plant or delivery of the batched overlay material would be more appropriate. The consistency of the material was such that it was difficult to move it around manually with racks and shovels as typically done with concrete overlays. This was demanding on the construction crew and thus more people were required to work with the material in front of the screed. Lightly misting the UHPC in front of the vibrating screed seemed to help with the consolidation of the layer and prevented the material from sticking as much to the vibrating ruler. Nonetheless, the crew placing and screeding the surface was about twice that required for a normal overlay job. One of the goals of this project was to place the UHPC overlay with basic concreting equipment, and the local contractor accomplished this goal using a simple vibrating screed machine and establishing a manual cranking set-up to move the screed along the length of the bridge. Whereas this process was satisfactory and cost effective, specially designed equipment for placing the UHPC overlay may be appropriate to place the material efficiently and provide a higher quality finish surface. It was felt that the screed ruler used for the Mud Creek Bridge project was neither taking full advantage of the UHPC thixotropic property nor sufficiently vibrating the material, making the top surface slightly rough. This required grinding of the surface prior to grooving.

The use of a curing compound was extremely important to keep the moisture within the UHPC layer immediately after consolidation by the screed machine. Determining the optimal curing time before grinding will need to be studied going forward so that this step can be completed cost effectively. Although 11.6 ksi was used as a threshold value for allowing the grinding to take place, it was felt that this threshold value is conservative and a lower value may be appropriate, which will reduce the duration of the project and facilitate grinding and/or grooving the surface more cost effectively. Grinding and grooving the surface on this project required relatively more effort because the UHPC was stronger and harder than a conventional overlay, but these steps were successfully carried out by a local contractor.

There was no special consideration given to the cold joint between the two overlay passes. A simple vertical joint was used. To ensure proper waterproofing of this joint, a key and/or a textured surface and even the use of short rebars across the joint may be needed to avoid water penetration in the long run. With further investigation over a period of years, the applicability of a simple vertical joint along the bridge length needs to be confirmed. While more intricate joint details may be more suitable, it will increase the cost of the overlay installation. Finally, the UHPC mixers required a three-phase power supply at the project site to operate, and the cost of this should be integrated into the contract upfront.

Based on this project experience and subsequent data gathering, the cost of the UHPC overlay material, including batching equipment used for this specific project, was in the range of US$20/ft^2. Similarly, the installation cost including grinding of the deck surface was in the same price range. At this stage, the total cost of UHPC overlay is therefore estimated to be about US$45/ft^2. Given that the Mud Creek Bridge was a smaller project with the first attempt to use UHPC overlay using a local crew and existing equipment, the cost per sq. ft for UHPC overlay is expected be lower as fixed costs are optimized with more field applications. Additional cost savings may be realized if the quality of placement of UHPC overlay mix can be achieved using appropriate equipment that can take advantage of its thixotropic property and provide a smooth finish surface ready for grooving.

Conclusions

This paper has presented a summary of steps taken to utilize UHPC as an overlay material for bridge deck surfaces in order to minimize the maintenance and repair costs of bridge decks in the U.S. Several steps were taken to: (a) establish a suitable UHPC mix formulation that had never been used before in North America; (b) verify its application under laboratory conditions; and (c) demonstrate placing of the material on an actual bridge deck.

All of the goals set out for this overlay project were successfully accomplished, which included the use of a local contractor, with no prior UHPC experience, to complete this small overlay project using basic concreting equipment and local manpower. The contractor used a larger crew than that required for a typical concrete overlay project and installed the UHPC overlay with a basic vibrating screed machine. Despite the success of this project, several improvements to the installation process have been identified. They include use of a small, special piece of equipment to place, vibrate and even groove the surface when placing the UHPC, thereby increasing the efficiency of batching the UHPC on site and reducing the construction time and cost.

The results from laboratory investigation were used as a guide to determine the surface roughness and thickness of the overlay. However, long-term monitoring of the Mud Creek Bridge deck is critical to ensure that the UHPC overlay increases the longevity of the bridge deck and reduces the maintenance cost. It is commonly believed that a UHPC overlay will be much more durable compared with various other overlay concepts used to date. However, more quantifiable benefits of the UHPC overlay are required to ensure that widespread application of the new technology would be economical and that it will help combat the widespread bridge deterioration problems facing the nation.

A 1.5 in. UHPC overlay is ideal as a running impermeable surface for a bridge deck. However, a UHPC layer in conjunction with a rebar mat can also be used to increase the deck capacity. Such a system could be used instead of performing full deck replacement. Using the remaining capacity of an existing deck and providing a top impermeable, durable UHPC surface would facilitate faster construction and be more sustainable than demolishing an existing deck and casting a completely new deck. This would take full advantage of UHPC as a high-performance material to build better systems.

The findings from the Iowa overlay and other related projects are unique and they can be used by other jurisdictions across North America to develop specifications assist with their use of UHPC in bridge applciations.

Acknowledgments

The research study and demonstration project presented in this paper were financially supported by the Federal Highway Administration, Iowa Department of Transportation, and Iowa Highway Research Board. Special thanks to Logan Wells of Iowa Department of Transportation for his effort toward the overlay design. In-kind contribution was also provided by Buchanan County Secondary Road Department and LarfargeHolcim.

References

1. Gucunski, N., F. Romero, S. Kruschwitz, R. Feldmann, and H. Parvardeh. *Comprehensive Bridge Deck Deterioration Mapping of Nine Bridges by Nondestructive Evaluation Technologies*. Final Report. Project no. SPR-NDEB(90)–8H-00, Iowa Highway Research Board, Ames, Iowa, 2011.
2. Aaleti, S., B. Petersen, and S. Sritharan. *Design Guide for Precast Ultra-High Performance Concrete Waffle Deck Panel System, including Connections*. Publication no. FHWA-HIF-13-032, Federal Highway Administration, U.S. Department of Transportation, 2013.

3. Aaleti, S., and S. Sritharan. Design of Ultrahigh-Performance Concrete Waffle Deck for Accelerated Bridge Construction. *Transportation Research Record: Journal of the Transportation Research Board*, 2014. 2406: 12–22.
4. Gheitanbaf, E. H., S. Sritharan, J. M. Rouse, and S. Aaleti. Bridge Decks with Precast UHPC Waffle Panels: A Field Evaluation and Design Optimization. *ASCE Journal of Bridge Engineering*, Vol. 21, No. 1, 2016.
5. Aaleti, S., S. Sritharan, and A. Abu-Hawash. Innovative UHPC-Normal Concrete Composite Bridge Deck. *Proc., RILEM-fib-AFGC International Symposium on Ultra-High Performance Reinforced Concrete*, (F. Toutlemonde and J. Resplendino, eds.), Rilem Publications SARL, Bagneux, France, 2013, pp. 217–226.
6. Brühwiler, E., M. Bastien-Masse, B. Houriet, B. Fleury, S. Cuennet, P. Schär, F. Boudry, and M. Maurer. Strengthening the Chillon Viaducts Deck Slabs with Reinforced UHPFRC. *Proc., of the IABSE Conference – Structural Engineering: Providing Solutions to Global Challenges*, Geneva, Switzerland, 2015.
7. *MCS-EPFL, Ultra-High Performance Fibre Reinforced Cement-Based Composites (UHPFRC): Construction Material, Dimensioning and Application.* English translation of the Technical Leaflet SIA 2052 with Adaptations (in French), Lausanne, Switzerland, 2016.
8. Sritharan, S. Design of UHPC Structural Members: Lessons Learned and ASTM Test Requirements. *Advances in Civil Engineering Materials, An ASTM International Journal*, Vol. 4, No. 2, 2015, pp. 113–131.
9. Bernardi, S., D. Jacomo, and F. Boudry. Overlay Ductal®: A Durable Solution for Bridges Retrofitting. *Proceedings of the First International Interactive Symposium on UHPC*. Des Moines, Iowa, 2016.

The Standing Committee on Construction of Bridges and Structures (AFH40) peer-reviewed this paper (18-06205).

Article

TRR
JOURNAL OF THE TRANSPORTATION RESEARCH BOARD

Transportation Research Record
2018, Vol. 2672(26) 48–56

DOI: 10.1177/0361198118758285
journals.sagepub.com/home/trr
SAGE

Leveraging Intelligent Compaction and Thermal Profiling Technologies to Improve Asphalt Pavement Construction Quality: A Case Study

George K. Chang[1], Kiran Mohanraj[1], William A. Stone[2], Daniel J. Oesch[2], and Victor (Lee) Gallivan[3]

Abstract

Intelligent compaction (IC) is an emerging technology with rollers equipped with global navigation satellite system (GNSS), an accelerometer-based measurement system, and an onboard color-coded display for real-time monitoring and compaction control. Paver-mounted thermal profiling (PMTP) is used to monitor asphalt surface temperatures behind a paver with a thermal scanner, and to track paver speeds, stops, and stop durations. Leveraging both IC and PMTP technologies allows for paving and compaction controls in real time, and for executing appropriate adjustments as needed. A case study is used to demonstrate the advantage of using both IC and PMTP over conventional operations. Postconstruction asphalt coring and tests, as well as pavement profile surveys were conducted to provide asphalt density data and pavement smoothness acceptance data for comparison and correlation analysis with IC and PMTP data. The data from 2 days of operations, one without the Material Transfer Vehicle (MTV) and another with the MTV, were analyzed and compared to illustrate the benefits of using IC, PMTP, and MTV for producing quality pavement products.

Durability and smoothness are two key construction qualities for agencies and users of hot mix asphalt (HMA) pavements. These two factors also affect the long-term structural and functional pavement performance.

The durability of an HMA pavement may typically be measured by the density achieved during construction, which is influenced by many factors such as mix type and design, nominal aggregate size to layer thickness ratio, mix production, mix temperatures, paver used, roller used, rolling amplitude/frequency, and roller passes, and so on. The current practice for density acceptance is normally measurement by a nuclear density gauge in the field or laboratory testing of cores extracted from the finished pavement. The limitations associated with these methods include: (a) a limited testing footprint when considering the entire pavement area; (b) under- or overcompaction with inadequate roller passes; and (c) non-uniform compaction caused by temperature segregation behind the paver. The smoothness of a pavement is normally determined using the international roughness index (IRI) measured by an inertial profiler. The smoothness of a finished pavement surface is affected by existing surface defects, overlay mix design, paver operations (especially paver stops), ski types, compaction, and so on.

Intelligent Compaction (IC) is an emerging technology with rollers equipped with global navigation satellite system (GNSS), an accelerometer-based measurement system, and an onboard color-coded display of various data maps for real-time monitoring and compaction control. IC allows the roller operator, quality control personnel, and agency to monitor roller passes and mat temperature, and so on (*1–4*). IC is standardized in the AASHTO PP81-17 specification (*5*) and required in more than 25 U.S. state department of transportation (DOT) construction specifications and special provisions (*6*).

Paver-mounted thermal profiling (PMTP) is used to monitor asphalt surface temperatures behind a paver with a thermal scanner, and to track paver speeds, stops, and stop durations. PMTP allows the screed operators, contractors, and agencies to monitor and review the uniformity and consistency of laydown temperatures (i.e., temperature differential or segregation) (*7, 8*). In general, higher temperature differentials result in lower pavement density caused by cooler areas of HMA. There are empirical relationships: (a)

[1]The Transtec Group, Inc., Austin, TX
[2]Missouri Department of Transportation, Jefferson City, MO
[3]Gallivan Consulting Inc., Carmel, IN

Corresponding Author:
Address correspondence to George K. Chang:
GKChang@TheTranstecGroup.com

25 °F temperature differential results in a 1% to 2% decrease in asphalt densities, and (b) 1% decrease in asphalt density results in a minimum of 10% reduction in pavement life (*9*, *10*). PMTP is standardized in the AASHTO PP80-17 specification (*11*) and required in some of U.S. DOT construction specifications.

Leveraging both IC and PMTP technologies allows for paving and compaction controls in real time, and for executing appropriate adjustments as needed. A case study of a Missouri Department of Transportation (MoDOT) state route project demonstrates the advantage of using both IC and PMTP over conventional operations. Postconstruction asphalt cores and tests, as well as pavement profile surveys were conducted to provide asphalt density data and pavement smoothness acceptance data for comparison and correlation analysis with IC and PMTP data. The data from 2 days of operations, one without the material transfer vehicle (MTV) and another with the MTV, were analyzed and compared to illustrate the benefits of using IC, PMTP, and MTV for producing quality pavement products.

Methodologies

Intelligent Compaction

IC technology allows real-time tracking of roller passes, surface temperatures, roller speeds, vibration amplitudes/frequencies, and intelligent compaction measurement value (ICMV). ICMV is the generic term for an accelerometer-based measurement that relates to stiffness of compacted materials (*1*). An IC system consists of the following components:

- Real-time kinematics (RTK) GNSS radio and receivers to provide precision positioning of roller drums, thus allowing tracking of roller passes;
- Accelerometers mounted on a non-dampened part of the drum to monitor responses from the compacted materials and underlying support layers;
- Non-contact temperature instrumentation to monitor the surface temperature of HMA material at the roller;
- Onboard display with color-coded maps of the above measurements.

Paver-Mounted Thermal Profiling

PMTP uses an infrared (IR) temperature scanner to continuously measure thermal profiles of mat behind a paver and allow monitoring of mat temperatures in real time. This allows detection of any thermal segregation and just-in-time adjustments if necessary. A PMTP consists of the following components:

- An IR scanner or camera for surface temperature measurement of the HMA behind the paver screed;
- GNSS antenna to provide the locations of scanned thermal profiles;
- An onboard computer to provide real-time visualization of mat temperatures behind the paver screed.

Pavement Profiling

Smoothness acceptance based on IRI requires true profiles be collected using either lightweight or highspeed inertial profilers. Such modern-day profilers are equipped with a laser height sensor, an accelerometer, and a distance measuring instrument (DMI) to collect profile traces in elevation for both left and right wheel paths. For use in the U.S., these profilers need to be certified based on AASHTO R56 standard (*12*) to ensure meeting the repeatability and accuracy standards for IRI analysis as per AASHTO R54 (*13*).

A Case Study

Project Description and Paving Operation

The project included placement of a new 1.75 in. HMA surface which overlaid the existing asphalt surface of the state route in Missouri. The approximate length of the project was 13 mi.

The first 2 days of paving operations at this project on April 24 and 25, 2017 (as Day 1 and Day 2, respectively) were considered for this case study. The paving equipment consisted of a CAT AP1055F paver (Figure 1*a*), a CAT CB64 IC roller for breakdown compaction, and another Caterpillar double-drum roller for finish rolling operations. A Roadtec SB2500 MTV (Figure 1*b*) was used on Day 2 of the paving operations. A MOBA Pave-IR PMTP was used at this project. The mounting of PMTP, thermal scanner, and onboard display are shown in Figure 2.

A Caterpillar double-drum IC roller was used as the breakdown roller. The IC-instrumented roller and IC data management system, VisionLink, are shown in Figure 3. The ICMV used was compaction meter value (CMV) (*1*). The technical specification of the IC roller is described in Table 1.

The coordinate reference system for all GPS receiver devices onsite was UTM-15N. MoDOT's state wide virtual reference system was used as the correctional signal. A Trimble GPS receiver was used for the IC roller. A Trimble hand-held rover was used to verify the accuracy of GPS receivers on IC rollers, and to measure the daily boundary points as well as spot test locations.

The paving mixture was a MoDOT specified BP-1, a 0.5 in. nominal maximum aggregate size 35 blow Marshall mix with PG 58-28 asphalt binder. The mix also consisted of 35% recycled asphalt pavements (RAP) and 1% recycled asphalt shingle (RAS), with a total effective binder replacement of 34%. The paving activities are summarized in Table 2.

A trial section test was conducted on Day 1 within the initial 1500 ft of the paving area. A nuclear density gauge

Figure 1. Paving operation: (*a*) CAT AP1055F paver and (*b*) Roadtec SB2500 MTV.

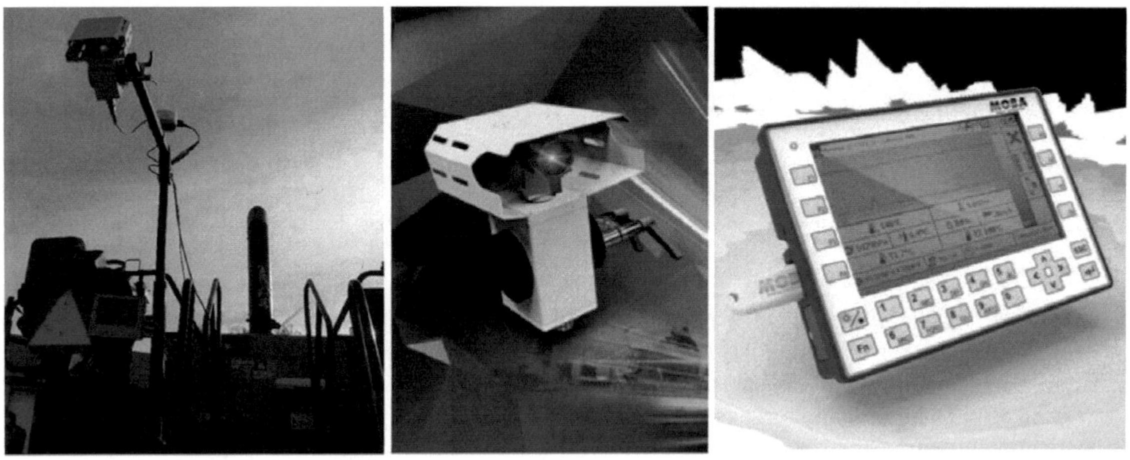

Figure 2. A Paver-Mounted Thermal Profiler system: (*a*) MOBA PAVE-IR mounted on the paver, (*b*) thermal scanner, and (*c*) onboard display.

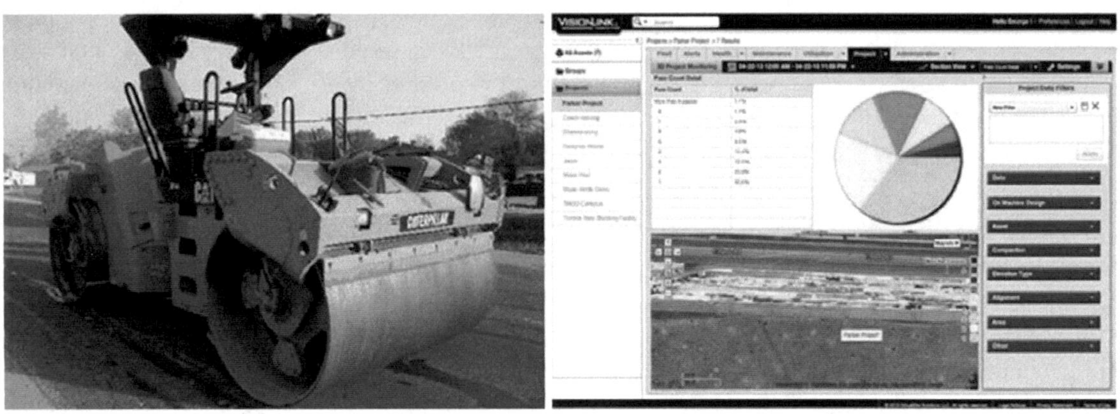

Figure 3. An IC system: (*a*) Caterpillar double-drum IC roller and (*b*) VisionLink data management system.

was used to perform the spot tests as per MoDOT specification procedures. The gauge was intended to measure density for 2 in. lifts (versus the actual 1.75 in. lift thickness). The compaction curve was established based on the density

Table 1. Technical Specification of the IC Roller

Parameter	Value
Manufacturer	Caterpillar
Model name	Tandem vibratory rollers
Model number	CB64B
Drum width	7 ft 8 in.
Machine weight	Operating weight 30,380 lb
Amplitude settings	0.016–0.041 in.
Frequency settings	2,520 and 3,800 vibrations per minute
ICMV	Compaction meter value (CMV) [unitless]
Documentation	VisionLink

thermal profile data with a given 150 ft sublot. The areas of any paver stop, 2 ft before and 8 ft after, were excluded from temperature differential computation as per AASHTO 80-17 specification. The classification of temperature differential is based on the range value as follows: low (range ≤ 25.0 °F); moderate (25.0 °F < range ≤ 50.0 °F); and severe (range > 50.0 °F).

The thermal profile analysis results are summarized in Table 3. There were significant changes in temperature segregation noticed between Day 1 and Day 2. The minor differences between the MOBA PPM and Veta results were caused by slight differences in computation methods.

Table 2. Paving Schedule and Activities

Day	Date	Direction/Lane	Start STN	StopSTN	Length (ft)	Lift	Width (ft)	Thickness (in)	HMA (tons)
1	4/24/2017	EB/L1	631+22	733+75.5	10,254	1	12	1.75	1,538
		EB/L1	740+16	758+55	1,839	1	12	1.75	
2	4/25/2017	WB/L1	631+22	733+78.5	10,257	1	12	1.75	1,910
		WB/L1	740+16	758+55	1,839	1	12	1.75	

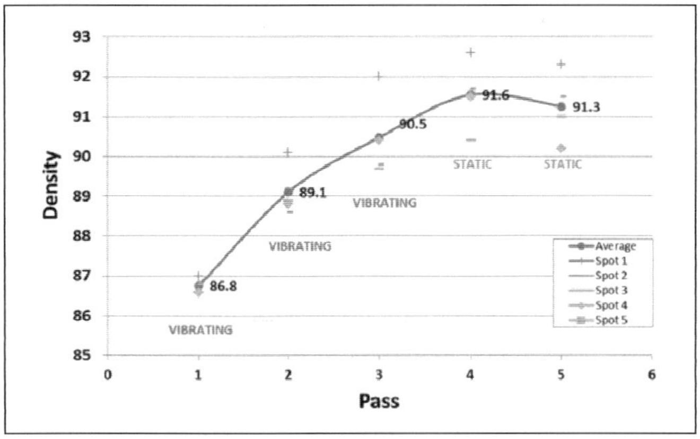

Figure 4. Compaction curve from a trial section (Day 1).

versus roller pass values as shown in Figure 4. Based on the compaction curve, the rolling pattern for the breakdown IC roller was determined to be three vibratory passes followed with one static pass. The vibratory passes were set to operate at high frequency and low amplitude.

Thermal Profiles Data Analysis

The MOBA PAVE-IR raw data was downloaded from a Cloud-based eRoutes data management system and analyzed with both MOBA Pave Project Manager (PPM) software and Veta software. The Veta software is an AASHTO standard software for IC-PMTP analysis developed by the first author (*6*). Veta uses the AASHTO 80-17 method to compute the "Range" values by taking the differences between the 98-percentile value and 1-percentile value of

IC Data Analysis

The IC data was downloaded from a Cloud-based VisionLink data management system and imported into Veta for analysis. The IC data analysis was performed based on the optimum roller pattern as determined by the trial section data. The "roller coverage" for each day of paving was classified according to the percentage of paved area which met or exceeded the optimum number of rolling passes based on the MoDOT specification: pass (roller coverage ≥ 90%); moderate (70% ≤ roller coverage < 90%); deficient (roller coverage < 70%). The target ICMV coverage is based on MoDOT IC specification: not flagged (ICMV coverage ≥ 70%); flagged (ICMV coverage < 70%). The mat temperature requirement is ≥ 225 °F for the vibratory passes. Both the target passes and target ICMV coverage requirements were passed, but the mat temperature for vibratory passes failed the requirements (Table 4). Note that this latter requirement was originally based on spot measurements instead of 100% coverage data such as IC.

Pavement Smoothness Analysis

The pavement smoothness data is shown below for Day 1 only. The pavement smoothness analysis was performed using the FHWA Profile Viewing and Analysis (ProVAL) software. The ProVAL software is an AASHTO standard software for pavement smoothness developed by the first author (*14*). The ride quality mean IRI (MRI) report was produced at a fixed interval of 528 ft (Figure 5) and the localized roughness report was a moving average of IRI at 25 ft (Figure 6) (*8*). In the fixed interval MRI report, all areas are above the

Table 3. Summary of Thermal Segregation Analysis Results

		MOBA PPM						Veta					
		Low		Moderate		Severe		Low		Moderate		Severe	
Day	Date	#	%	#	%	#	%	#	%	#	%	#	%
1	4/24/2017	0	0	3	3	87	97	0	0	1	1	77	99
2	4/25/2017	17	15	87	80	5	5	16	20	61	76	3	4

Table 4. Summary of IC Analysis Results for Coverages

		Target passes		Target ICMV		Temperature for vibratory passes	
No.	Date	% Coverage	Classification	% Coverage	Classification	% Coverage	Classification
1	4/24/2017	95.5	Passed	92.25	Pass	51.37	Failed
2	4/25/2017	97.7	Passed	91.45	Pass	47.55	Failed

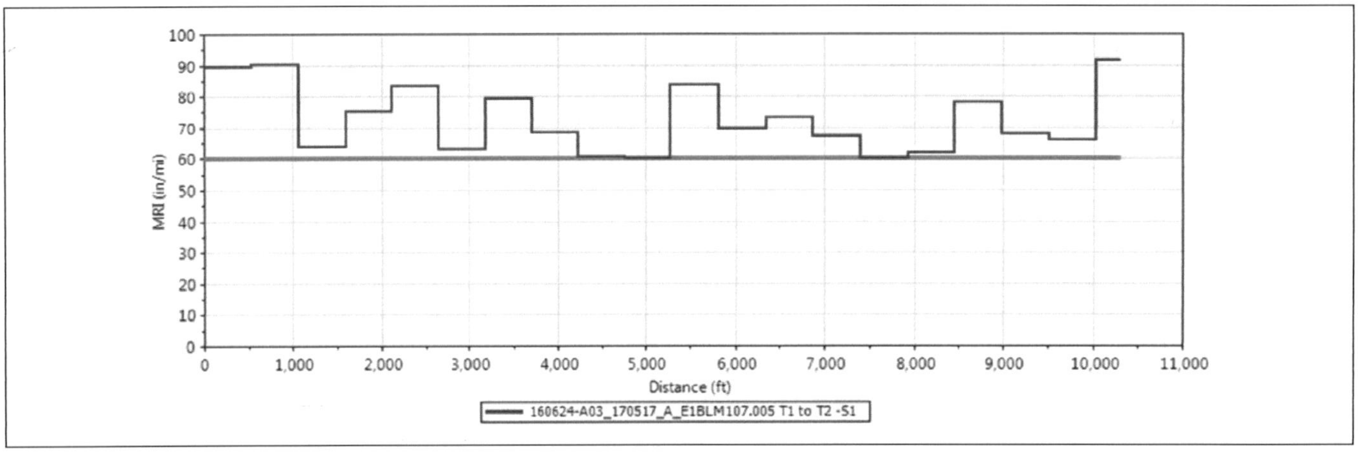

Figure 5. Fixed interval MRI report (Day 1).

acceptance value of 60 in./mi. There are also several areas of localized roughness (ALR) that exceeds 125 in./mi.

Correlation and Comparison Analysis

The following includes the correlation analysis and comparison of the above test results.

Comparison of Temperature Segregation between Day 1 and Day 2. The temperature segregation results from the Veta analysis for Day 1 and Day 2 (Figure 7) are significantly different. With the use of the MTV, severe temperature segregation drops from 99% to 4% on Day 1 and Day 2, respectively. This is an excellent control case because all other factors are equal between the 2 days of paving except the use of the MTV.

Comparison of Localized Roughness between Day 1 and Day 2. The ALRs drop from Day 1 to Day 2, though not as significantly as the temperature differential (Figure 8). The use of the MTV reduces temperature segregation and indirectly helps improve the smoothness.

Comparison of CMV/Passes and Asphalt Core Density on Day 1. The correlation of the final coverage CMV and core density for Day 1 is shown in Figure 9. The R^2 coefficient of linear correlation is 0.62, which is greater than the generally accepted value of 0.5 in most IC specifications. Therefore, it is valid to use the linear correlation to calculate the target CMV corresponding to 92% of G_{mm} maximum theoretical specific gravity for the core density data. The resulting target CMV is 10.68 (unitless). The CMV value for the failure criteria can also be computed based on the formula mean − 1.28*STD (where STD denotes standard deviation), resulting in 4.64 units. The resulting value corresponds to the 90-percentile statistic, and was used to determine whether the CMV coverage meets the target for Day 1. The core densities from Day 1 to Day 2 improve

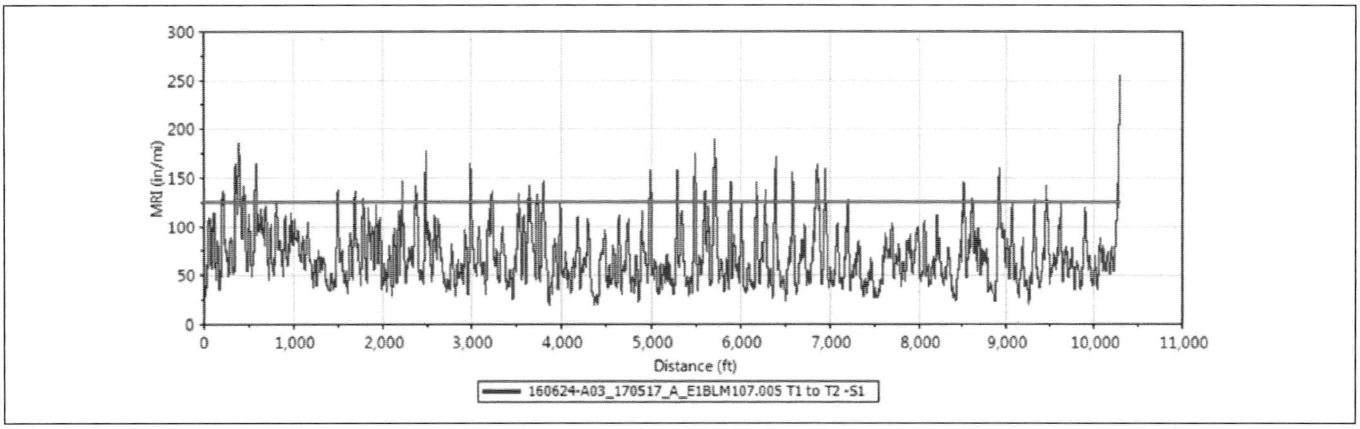

Figure 6. Localized roughness MRI report (Day 1).

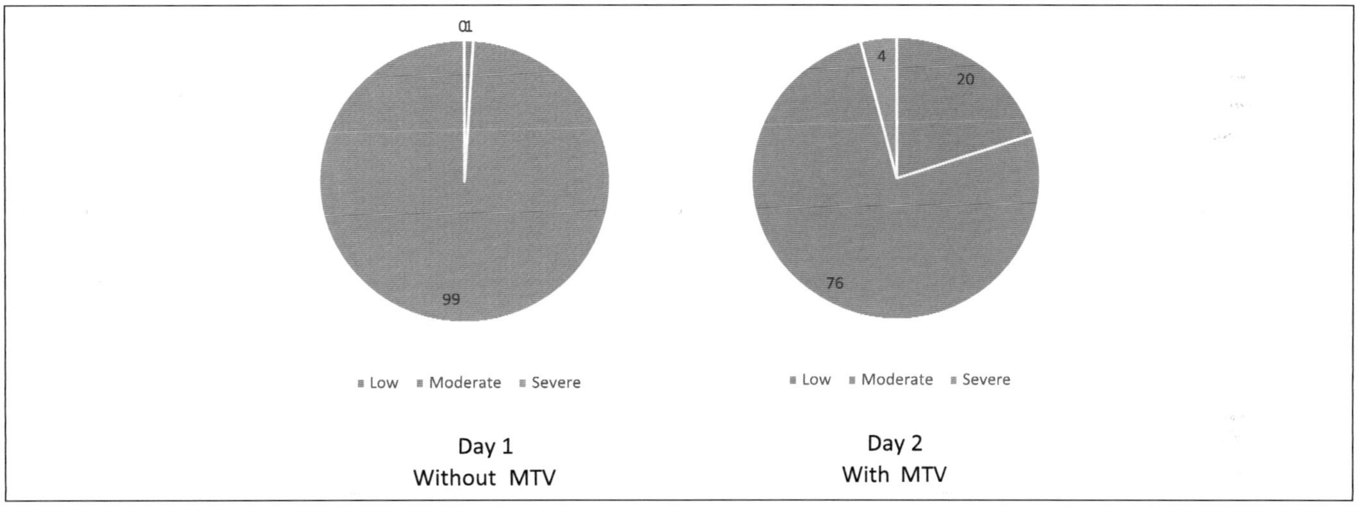

Figure 7. Comparison of temperature segregation between Day 1 (without MTV) and Day 2 (with MTV).

slightly (average: 94.7–95.5, STD: 1.0–0.7) based on limited four core data from each day.

The correlation between pass count and core density is excellent with $R^2 = 0.96$, indicating a well-designed and well-produced mix that can be compacted with an adequate roller at the appropriate frequency/amplitude/temperature (shown in Figure 10).

Comparison of Paver Stops and Localized Roughness on Day 1. The relatively colder mat temperatures coincide with paver stop locations for most cases. The paver stop locations also mirror the ALRs, as shown in Figure 11. This is a classic example of paving operation versus localized roughness. Therefore, any paver stop location would often result in temperature segregation, undesirable asphalt density, and localized roughness. Adequate planning and execution of paving operation is important to reduce paver stops and their associated issues.

The comparison of temperature differential (150-ft sublots), fixed interval MRI (150-ft sublots), and area of localized roughness is shown in Figure 12. The comparison indicates a similar trend among these three metrics.

Summary and Conclusions

An asphalt paving field project was conducted on 2 days to demonstrate the advantages of using both IC and PMTP technologies to improve construction quality control. An MTV was used on Day 2 only, with Day 1 as a control case. Postconstruction pavement profiling and asphalt core densities were also collected as acceptance tests. Based on the above data analysis, the following are highlighted observations made for this case study:

- Adequate roller coverage was achieved for both Day 1 and Day 2 with 95.5% and 97.7% of the pavement

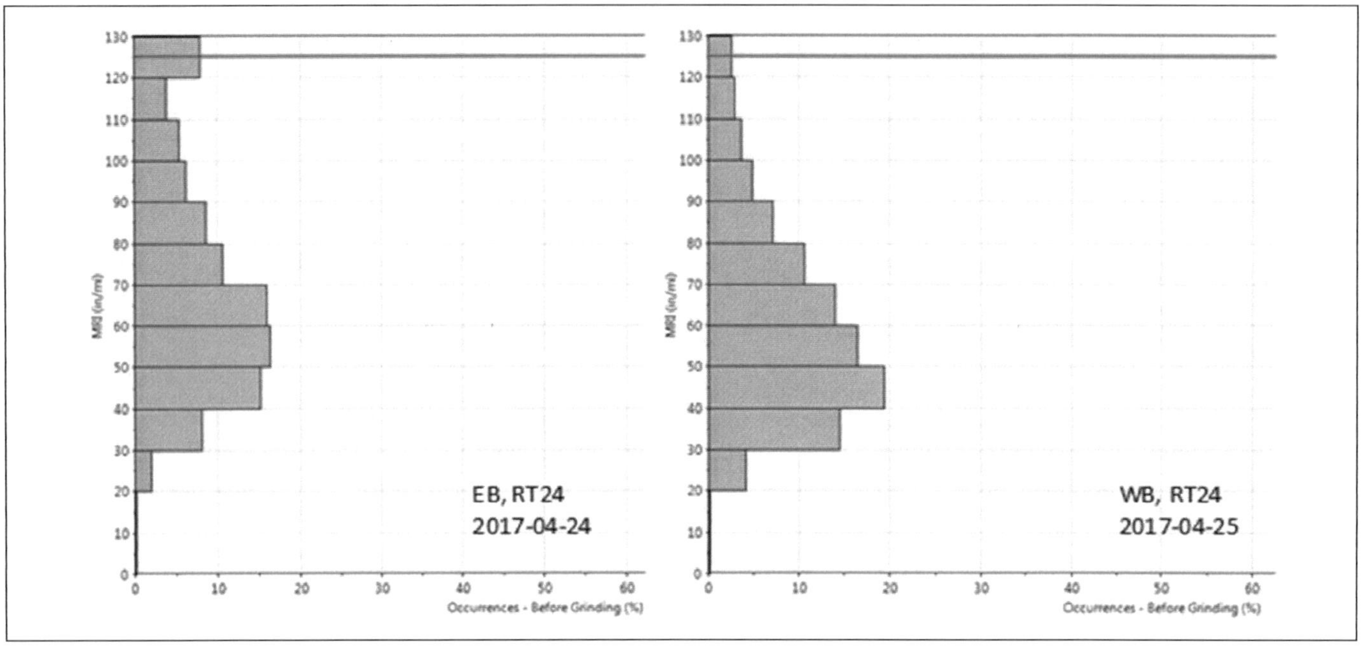

Figure 8. Comparison of localized roughness between Day 1 (without MTV) and Day 2 (with MTV).

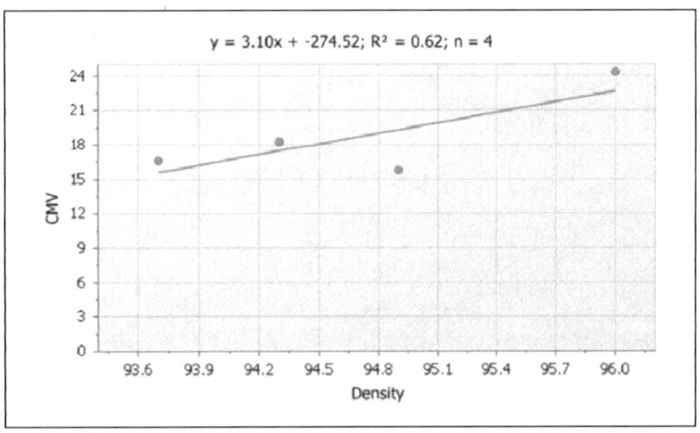

Figure 9. Correlation between ICMV and asphalt core densities (Day 1).

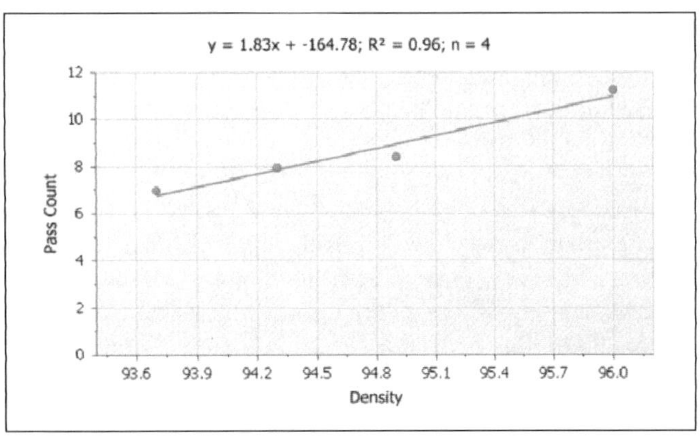

Figure 10. Correlation between roller passes and asphalt core densities (Day 1).

receiving the target number of passes or above respectively, indicating an effective use of the IC system.

- The PMTP temperature segregation results from Day 1 and Day 2 are significantly different. With the use of the MTV, temperature segregation in the Severe category changes from 99% on Day 1 to 4% on Day 2.
- The ALR from pavement profile data changes slightly from Day 1 to Day 2. There were a similar number of paver stops for both days.
- The correlation of the final coverage ICMV and core density is with $R^2 = 0.62$, considered as passing in most IC specifications. Therefore, the correlation may be used to determine target ICMV for quality control.
- The correlation between pass count and core density is with $R^2 = 0.96$. This indicates a well-designed and well-produced mix that can be compacted with an adequate roller at the appropriate frequency/amplitude/temperature.
- Locations of colder mat temperatures are consistent with those at paver stops. Paver stop locations also mirror ALRs.
- The metrics of temperature differential (150-ft sublots), fixed interval MRI (150-ft sublots), and area of localized roughness indicate a similar trend.

Using PMTP, IC, and pavement profiling technologies, the entire paving process can be monitored, as demonstrated in this case study: (a) PMTP collects continuous measurements of thermal profiles of the asphalt mat behind the paver; (b) IC tracks roller passes, asphalt surface temperatures for each

Figure 11. Paver stops/thermal profiles (top) and localized roughness MRI report (bottom) (Day 1).

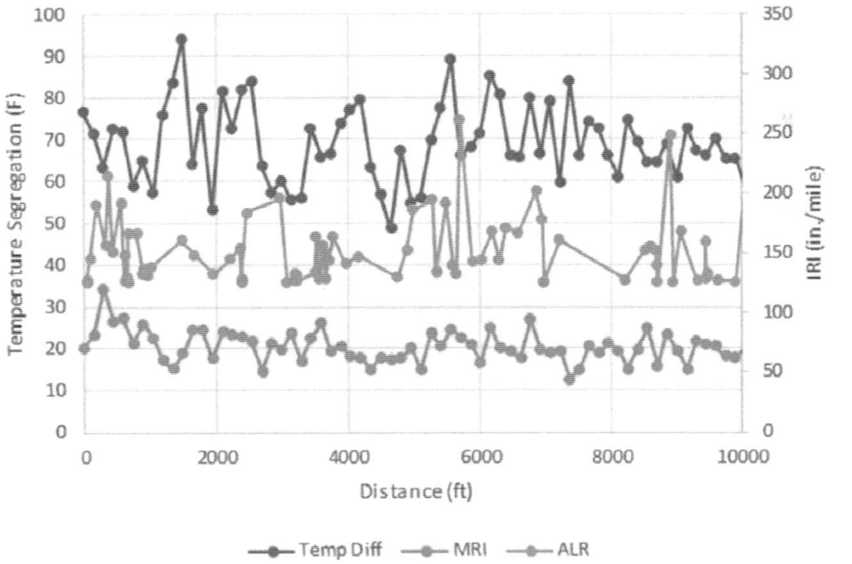

Figure 12. Comparison of temperature differentials, fixed interval MRI report, and area of localized roughness MRI report (Day 1).

pass, ICMVs, roller vibration frequencies and amplitudes; (c) Pavement profiling measures the smoothness of the finished surfaces along the wheel paths. As demonstrated in this case study and other similar studies, leveraging both IC and PMTP technologies allows for paving and compaction controls in real time, and for executing appropriate adjustments as needed. By combining the above three technologies, construction issues can be readily identified to improve pavement quality and to make pavements last longer.

Acknowledgments

The authors graciously thank the FHWA Accelerated Innovation Deployment (AID) program and MoDOT for funding this IC and PMTP project. This paper does not endorse any commercial products. The state route and contractor's name for the case study were concealed for privacy.

References

1. Chang, G. K., Q. Xu, J. Rutledge, B. Horan, L. Michael, D. White, and P. Vennapusa. *Accelerated Implementation of Intelligent Compaction Technology for Embankment Subgrade Soils, Aggregate Base, and Asphalt Pavement Materials*. FHWA-IF-12-002. FHWA, U.S. Department of Transportation, 2012.
2. Chang, G. K., V. L. Gallivan, and Q. Xu. Assess Asphalt In-Place Density with Intelligent Compaction Measurements. In *Asphalt Pavements* (Y. R. Kim, ed.), Boca Raton, FL, CRC Press, 2014, pp. 489–514.
3. Xu, Q., and G. K. Chang. Experimental and Numerical Study of Asphalt Material Geospatial Heterogeneity with Intelligent Compaction Technology on Roads. *Journal of Construction & Building Materials*, Vol. 72, 2014, pp. 189–198.
4. Xu, Q., and G. K. Chang. Adaptive Quality Control and Acceptance of Pavement Material Density for Intelligent Road Construction. *Journal of Automation in Construction*, Vol. 62, 2016, pp. 78–88.
5. AASHTO PP81-17. *Standard Practice for Intelligent Compaction Technology for Embankment and Asphalt Pavement Applications*. American Association of State and Highway Transportation Officials, 2017.
6. *Intelligent Compaction Website*. http://www.intelligentcompaction.com/. Accessed November 2, 2017.
7. Sebesta, S, and T. Scullion. *Statewide Implementation of Pave-IR in the Texas Department of Transportation*. TTI 5-4577-05-1. Texas Department of Transportation, 2012.
8. Landefeld, C. *Evaluation of Pave-IR Thermal Imaging for Asphalt Pavement Uniformity*. SJN 49509. Ohio Department of Transportation, 2014.
9. Willoughby, K. A., J. P. Mahoney, L. M. Pierce, J. S. Uhlmeyer, K. W. Anderson, S. A. Read, and R. Moore. Construction-Related Asphalt Concrete Pavement Temperature Differentials and the Corresponding Density Differentials. *Research Report, Research Project Agreement T9903, Task A3, Cyclic Segregation*. Washington State Transportation Commission, Olympia, Washington, 2001.
10. Sebesta, S., and T. Scullion. *Performance Monitoring Pavements with Thermal Segregation in Texas*. FHWA/TX-12/0-6080-1. FHWA, U.S. Department of Transportation, 2012.
11. AASHTO PP80-17. *Standard Practice for Continuous Thermal Profile of Asphalt Mixture Construction*. American Association of State and Highway Transportation Officials, 2017.
12. AASHTO R56-14. *Standard Practice for Certification of Inertial Profiling Systems*. American Association of State and Highway Transportation Officials, 2014.
13. AASHTO R54-14. *Standard Practice for Accepting Pavement Ride Quality When Measured Using Inertial Profiling Systems*. American Association of State and Highway Transportation Officials, 2014.
14. *Profile Viewing and Analysis software (ProVAL)*. www.RoadProfile.com. Accessed November 2, 2017.

The Construction Section (AFH00) peer-reviewed this paper (18-01766).

Article

TRR
JOURNAL OF THE TRANSPORTATION RESEARCH BOARD

Transportation Research Record
2018, Vol. 2672(26) 57–67

DOI: 10.1177/0361198118767419
journals.sagepub.com/home/trr
SAGE

The FHWA's Demonstration Project for Enhanced Durability of Asphalt Pavements through Increased In-Place Pavement Density

Tim Aschenbrener[1], E. Ray Brown[2], Nam Tran[2], and Phillip B. Blankenship[3]

Abstract

Recognizing the importance of in-place density in building cost-effective asphalt pavements, a Federal Highway Administration (FHWA) Demonstration Project was created for "Enhanced Durability of Asphalt Pavements through Increased In-Place Pavement Density." The objective of the demonstration project was to determine the benefit of additional compaction and show that additional density could be obtained through improved techniques. This project effort included two major components: (1) a literature search to serve as an educational component regarding the best practices for increasing density, and (2) the construction of 10 field demonstration projects. Eight of the 10 states improved densities by at least 1% compared to a control section on their field demonstration projects. There were at least two pavement sections (a control and at least one test section) constructed within each of the 10 states that participated in this field demonstration project. Many of the states constructed more than two pavement sections for a total of 38 sections. There were many variables, including mixture type, construction equipment, and procedures between states and within states. A summary of the methods that states used to obtain increased density generally fell into one of five categories: (1) improving the agency's specification by including or increasing incentives and increasing the minimum percentage density requirements; (2) making engineering adjustments to the asphalt mixture design to obtain slightly higher optimum asphalt content (although not part of the original goal of the demonstration project); (3) improving consistency as measured by the standard deviation; (4) following best practices; and (5) using new technologies.

In-place density is one of the most important factors that can significantly influence the long-term performance of an asphalt pavement (*1*). A small increase in in-place density can potentially lead to a significant increase in the pavement service life. Based on past studies reviewed in a previous report (*2*), a 1% decrease in in-place air voids was estimated to improve the fatigue performance of asphalt pavements between 8 and 44% and improve rutting resistance by 7–66%. In addition, based on field data, a 1% decrease in in-place air voids would extend the service life by 10%, conservatively.

To illustrate the effect of in-place density on the life cycle cost of asphalt pavements, a life cycle cost analysis (LCCA) was conducted on two pavements in which the same asphalt overlay would be constructed to 93% and 92% (densities) of the theoretical maximum density (TMD). Using the conservative 10% increase in service life, the LCCA results revealed that the state highway agency (SHA) would see cost savings in net present value (NPV) of $88,000 on a $1,000,000 paving project (8.8%) by increasing the minimum required density by 1% of TMD (*2*). This saving does not consider potential savings from other costs such as operation, maintenance, and road-user costs.

Recognizing the importance of in-place density in building cost-effective asphalt pavements, a Federal Highway Administration (FHWA) Demonstration Project was initiated for "Enhanced Durability of Asphalt Pavements through Increased In-Place Pavement Density." The Demonstration Project was the partnership between FHWA, SHAs, the National Asphalt Pavement Association, and the contractors that built the demonstration sections. This paper summarizes activities, observations, and lessons learned from the demonstration project.

[1]Federal Highway Administration, Lakewood, CO
[2]National Center for Asphalt Technology, Auburn, AL
[3]Blankenship Asphalt Tech and Training, Richmond, KY

Corresponding Author:
Address correspondence to Tim Aschenbrener: timothy.aschenbrener@dot.gov

Objective and Scope

The overall objective of the demonstration project was to achieve increased in-place density for asphalt pavements. There were two major components of this study, including (1) a literature search to serve as an educational component for the best practices for increasing in-place density; and (2) the construction of 10 field demonstration projects. The field demonstration projects were constructed in 10 states whose participation was selected through an application process. Each of the SHAs and contractors involved in the field demonstration projects were offered a workshop from the Asphalt Institute (AI) and FHWA and on-site assistance during construction from the National Center for Asphalt Technology (NCAT).

Each field demonstration project included a control and one or more test sections. The control section was built by the contractor to achieve the required in-place density based on their routine construction practices. At least one test section was built as part of the agreement with FHWA, and the goal of this section was to use improved paving and compaction techniques to increase density. The contractor was encouraged to employ techniques that did not require additional rollers or a higher asphalt content that would result in significantly increased cost. In some states, additional test sections were constructed by the SHAs and contractors to evaluate other techniques, which generally included additional rollers to improve density or other ideas of interest that they believed would work best in their local state.

Background

A literature search and review of SHA specifications was conducted to identify best practices and new technologies that could help achieve density. Higher in-place density can be obtained to improve the long-term performance of asphalt pavements in a cost-effective manner by adopting some of the practices and technologies briefly described in Table 1.

Field Demonstration Projects

There were at least two pavement sections constructed within each of the 10 states that participated in the demonstration project to enhance durability through increased in-place density. Several states constructed more than two pavement sections, with a total of 38 sections being constructed.

There were many variables including mixture types, construction equipment, and procedures between states and within states, making it very difficult to compare the density results between various pavement sections. The number of variables that were intentionally changed within a state was much less than the number of changes between states. This was expected, as it was a demonstration project and not a formal experiment. As a demonstration project, each state (the contractor and agency) was empowered to focus on changes to improve density that they thought would be most beneficial for their situation. Thus, it was much easier to compare the changes made within a state to show the effect of these changes on performance.

A summary of the asphalt mixture data along with in-place density relative to the TMD is provided in Table 2. The primary control sieve and control point as defined in AASHTO M 323 were used to make the determination of fine or coarse grading. Some mixture and construction attributes are further described in the following subsections. "Prod Air Voids" and "Prod VMA" are measured on plant-mixed, lab-compacted specimens at N_{design}.

Gradation Type

In-place density is related to permeability. Permeability is also impacted by the type of gradation (coarse versus fine) and the nominal maximum aggregate size (NMAS). A 1% improvement in density may have more impact on the long-term performance of a coarse gradation with a larger NMAS than a finer gradation with a smaller NMAS. The breakdown of gradations used by each state is as follows:

- Four states used fine gradations (states 1, 4, 5, and 10).
- Six states used coarse gradations (states 2, 6, 7, 8, 9, and 10).
- The gradations were not available for state 3, and state 10 used both fine-graded and coarse-graded mixtures.

Experience has shown that fine-graded mixtures are generally more workable and easier to compact than coarse-graded mixtures. It is clear from the data in Table 2 that good or poor density could be obtained with either fine-graded or coarse-graded mixtures. Based on these data, it appeared that rolling procedures could generally be adjusted to obtain adequate density when mixture variables such as air voids, NMAS, and laboratory compaction level were varied. There were many other factors, such as mixture volumetric properties, that likely had a greater effect on in-place density than the gradation type.

Nominal Maximum Aggregate Size

The breakdown of the NMAS used by each state is as follows:

- Four states used an asphalt mixture with 9.5-mm NMAS (states 3, 4, 6, and 7).
- Seven states used an asphalt mixture with 12.5-mm NMAS (states 1, 2, 3, 4, 5, 8, and 9).
- One state used an asphalt mixture with 19-mm NMAS (state 10).

Table 1. Summary of Practices and Technologies for Improving In-Place Density

Process	Practices and technologies
Mix design and field verification	• Fine-graded Superpave mixes can be used in place of coarse-graded Superpave mixes to improve field compaction without affecting the long-term performance of asphalt pavements (*3,4*). • During pavement design, the lift thickness should be designed to be a minimum of three and four times the intended nominal maximum aggregate size (NMAS) for fine- and coarse-graded mixes, respectively. The thicker the lift, the more room for compaction. Lift thickness is related to potential density, not to rutting (*5*). • For some SHAs, mix design requirements have been refined to encourage increasing effective binder volume. An example is provided with Superpave 5 (*6*). More guidance is provided in a tech brief by FHWA (*7*). Some of these requirements should only be used after local experience. These changes can improve field compaction while ensuring mixture resistance to premature distresses such as rutting, cracking, and moisture damage. • After a mix design is completed in the laboratory, it should be verified and properly adjusted at the start of production as materials in the field may be different and/or more variable than those used in the laboratory, and field-acceptance criteria may be different from those used for the asphalt mixture design.
Field compaction	• The underlying layers should be properly constructed and inspected to provide sufficient, consistent support for achieving higher in-place density. • Appropriate compaction equipment should be selected and properly operated during paving. The rolling pattern should be optimized to achieve both in-place density and consistency (*8,9*). Paving operations should be balanced to improve the ability to obtain density and consistency (*10*). • It is important to understand how weather conditions can affect the mix temperature. If needed, the MultiCool software can be used to estimate the available time for compaction (*11*).
Measurement and payment	• The in-place field density should be compared with TMD from field-produced samples. Useful information on the bulk specific gravity of the mixture (G_{mb}) and TMD is presented in a FHWA tech brief (*12*). • Incentive specifications can be adopted to yield higher in-place density. A good SHA specification should include an asphalt mixture design procedure that can result in workable and compactable mixtures with an incentive that is obtainable for in-place density (*13,14*). • Utilizing good specifications, the Pennsylvania Department of Transportation (PennDOT) and New York State Department of Transportation (NYSDOT) could obtain good in-place density results using the minimum lot average specification and the percentage within limits (PWL) specification, respectively (*15*).
New technologies	• Warm-mix asphalt (WMA) can be utilized to improve compaction, especially for projects requiring longer haul times and/or constructed in cold weather temperatures and conditions (*16*). • Intelligent compaction (IC) can be implemented to make it easier to optimize, automate, and monitor compaction parameters such as rolling pattern, frequency, drum spacing, amplitude, temperature, and number of coverages to achieve higher in-place density and consistency (*17,18*). • Infrared (IR) imaging can be deployed to measure the real-time mat temperature and adjustments made to improve temperature consistency and in-place density (*19*).
Others	• Best practices should be followed to achieve optimal compaction for longitudinal joints. The Asphalt Institute website (*20,21*) has more detailed information about specifying and constructing longitudinal joints. • Tack coats should be applied sufficiently and uniformly to improve compaction. A good tack coat application will assist compaction and provide an improved bond, resulting in better long-term performance (*1, 22*).

Note: TMD = theoretical maximum density.

Changing the NMAS also changed the thickness to NMAS (t/NMAS) when the layer thickness remained the same. This made it difficult to make a direct comparison between two different NMASs. Generally, it is desirable that the t/NMAS be at least 3.0 for fine-graded mixtures and at least 4.0 for coarse-graded mixtures. The t/NMAS used on the field demonstration projects generally followed best practice guidelines. The t/NMAS on the field demonstration projects were as follows:

- One of 10 states <3.0 (state 10).
- Nine of 10 states ≥3.0 (states 1, 2, 3, 4, 5, 6, 7, 8, and 9).
- Eight of 10 states ≥4.0 (states 1, 2, 3, 4, 5, 6, 7, and 9).

Table 2. Summary of Mixture Properties and In-Place Density

State and section number	NMAS (mm)	Fine-graded or coarse-graded	Thick to NMAS	Number of gyrations	Mix design AC (%)	Mix design air voids (%)	Prod air voids (%)	Mix design VMA (%)	Prod VMA (%)	Density (% of TMD)
1-C	12.5	Fine	4.0	100	5.0	4.0	3.7	14.1	13.7	93.5
1-TS1	12.5	Fine	4.0	100	5.0	4.0	3.3	14.1	13.3	93.2
1-TS2	12.5	Fine	4.0	100	5.0	4.0	3.3	14.1	13.4	95.4
2-C	12.5	Coarse	4.0	100	5.0	4.1	4.3	15.9	15.1	91.0
2-TS1	12.5	Coarse	4.0	100	5.0	4.1	4.3	15.9	15.1	91.8
3A-C	12.5	–	3.0	90	5.2	–	–	–	–	92.9
3A-TS1	12.5	–	3.0	90	5.2	–	–	–	–	92.9
3A-TS2	12.5	–	3.0	60	5.5	–	–	–	–	93.5
3A-TS3	12.5	–	3.0	60	5.5	–	–	–	–	94.1
3B-C	12.5	–	3.0	60	–	–	–	–	–	93.7
3B-TS1	12.5	–	3.0	60	–	–	–	–	–	94.9
3B-TS2	9.5	–	4.0	60	–	–	–	–	–	93.6
3B-TS3	9.5	–	4.0	60	–	–	–	–	–	94.2
3B-TS4	9.5	–	4.0	60	–	–	–	–	–	93.7
3B-TS5	12.5	–	3.0	60	–	–	–	–	–	93.8
4-C	12.5	Fine	3.5	75	5.5	4.0	4.6	15.8	15.9	93.5
4-TS1	12.5	Fine	3.5	75	5.5	4.0	4.4	15.8	15.7	95.0
4-TS2	12.5	Fine	3.5	75	5.8	3.0	3.4	15.8	15.3	94.6
4-TS3	12.5	Fine	3.5	75	5.8	3.0	2.6	15.8	14.7	95.4
4-TS4	12.5	Fine	3.5	75	5.5	4.0	4.4	15.8	16.0	92.5
4-TS5	12.5	Fine	3.5	75	5.8	3.0	3.8	15.8	15.6	93.4
4-TS6	12.5	Fine	3.5	75	5.5	4.0	3.2	15.8	15.5	94.0
4-TS7	9.5	Fine	4.7	75	5.7	4.0	3.8	15.7	16.0	95.2
5-C1	12.5	Fine	4.0	50	5.3	4.0	3.6	14.8	14.4	92.5
5-TS1	12.5	Fine	4.0	50	5.3	4.0	3.6	14.8	14.4	93.2
5-TS2	12.5	Fine	4.0	50	5.6	3.0	2.8	14.7	15.0	95.2
6-C	9.5	Coarse	4.0	100	6.7	4.0	4.7	15.2	15.3	93.3
6-TS1	9.5	Coarse	4.0	50	6.8	5.0	5.6	16.5	16.5	95.4
7-C1	9.5	Coarse	4.0	75	6.2	3.5	3.4	16.2	16.5	94.4
7-TS1	9.5	Coarse	4.0	75	6.2	3.5	3.4	16.2	16.5	96.1
8-C1	12.5	Coarse	3.6	100	5.7	5.5	3.5	16.4	14.6	93.1
8-T1	12.5	Coarse	3.6	100	5.7	5.5	3.5	16.4	14.6	93.0
9-C	12.5	Coarse	4.0	50	5.6	3.1	3.6	15.6	15.8	92.2
9-TS1	12.5	Coarse	4.0	50	5.6	3.1	2.7	15.6	15.3	92.0
9-TS2	12.5	Coarse	4.0	50	5.6	3.1	3.4	15.6	16.0	92.0
10-C	19.0	Fine	2.7	75	5.6	4.0	–	16.6	–	95.6
10-TS1	19.0	Coarse	2.7	75	5.6	4.0	–	16.6	–	95.0
10-TS2	19.0	Coarse	2.7	75	5.5	3.7	–	15.1	–	95.7

Note: NMAS = nominal maximum aggregate size; TMD = theoretical maximum density; VMA = voids in the mineral aaggregate.

States 3 and 4 both evaluated the effect of two different NMASs with the same lift thickness. Each state produced at least one section with a 12.5-mm mixture and at least one section with 9.5-mm mixture. State 3 showed that a 94.1% average density was obtained with a 12.5-mm mixture and 93.8% average density was obtained with a 9.5-mm mixture.

These density results were not significantly different. State 4 showed 94.1% average density for the 12.5-mm mixture and 95.2% density for the 9.5-mm mixture. This difference of 1.1% density is probably significant. The purpose of changing the NMAS was to examine the effect of the t/NMAS.

Placement and Compaction

The placement and compaction data along with in-place density results are provided in Table 3. A pass was defined as the roller passing over one point in the mat one time. Material transfer vehicles (MTVs) have been shown to provide improved smoothness and reduced segregation and were used on eight of the 10 projects. The states that used MTVs on at least one of the pavement sections were states 1, 4, 5, 6, 7, 8, 9, and 10.

Some states specified and controlled density using a method other than percentage of TMD. For Tables 2 and 3, the density is reported for all states as percentage of TMD. Two of the 10 states that used a method other than TMD were states 2 and 9.

The primary property that is important during compaction is the percentage of air voids in the in-place mixture. Reporting density as percentage of TMD directly provides the air voids in the compacted mix. Other methods of specifying and measuring density only provide an indirect measure of the air voids and, in some cases, can be misleading.

A summary of the percentage density in the most successful test section for each state is shown in Table 4. There was a significant difference in the number of passes between the various states. The two states with the lowest compaction effort had significantly lower density than that in the other states. The data are sorted by total passes.

Lessons Learned to Achieve Increased In-Place Density

In-place density can be improved through focused efforts on field compaction. Eight of 10 states improved densities by at least 1% on their field demonstration projects. One of these states did this by improving the consistency or standard deviation. There was enough improvement in the standard deviation to effectively raise the lower specification limit by 1%. Seven of the 10 states improved densities to a minimum of 94.0% of the TMD. Based on the observations from these 10 field demonstration projects, techniques were identified to improve density that will be of interest to agencies and contractors.

In-Place Density Requirement

Agency specifications play a key role in the amount of density obtained on a project. The contractor's job is to be the low bidder and meet the specifications. In two of 10 states (states 1 and 2), higher density was achieved simply by the agency asking for it. In state 1, an additional roller was added. In state 2, more passes were used by the existing roller. Although this would not work in all states, some states could simply raise the minimum density requirements and the contractors could easily adjust their compaction methods to meet specifications.

Optimum Asphalt Content

Some states obtain higher density by adding more asphalt binder to the mix and others obtain higher density by increasing compaction with more rollers. These two approaches of reducing the in-place air voids do not have the same effect on performance. It is important that a satisfactory mix be designed and produced to ensure good performance and that the mix be compacted to an adequate density in the field.

Four of 10 states, including states 3, 4, 5, and 6, made engineering adjustments to the AASHTO Superpave mix design to obtain higher optimum asphalt content. These states had an increase of 0.1 to 0.3% asphalt. Engineering adjustments to obtain a slightly higher optimum asphalt content included adjusting gyrations (states 3 and 6) and air void regression (states 4 and 5). The effects of these adjustments on the mixture performance properties have been evaluated in previous studies. A brief description of each adjustment follows.

- The gyration level for state 3 was varied, and in this case the increase in density was 1.2%. State 6 reduced the gyration level from 100 to 50 and the in-place density increased by approximately 2.1%. State 6 simultaneously increased the air void content at design from 4.0 to 5.0% and increased the voids in the mineral aggregate (VMA) requirement from 15.0 to 16.0%. Therefore, as expected, states 3 and 6 showed that lower gyrations during laboratory compaction resulted in a higher in-place density (fewer air voids).
- Another factor in mix design that influences density is the design air void level. A pavement section designed with lower design air voids will be easier to compact than one with higher design air voids for the same gradation. Two states that looked at varying the laboratory air voids without significantly changing other mixture properties or compaction procedures were states 4 and 5. The results from state 5 showed that lowering the design air voids from 4.0 to 3.0% resulted in an approximate 2.5% increase in in-place density. The results from state 4 showed that lowering the design air voids from 4.0 to 3.0% without changing the gradation resulted in an approximate 1.9% increase in in-place density.

For other states that wish to make these adjustments in the future to increase the binder content to a level that is higher than the optimum binder content provided in the AASHTO Superpave standards, it is recommended that performance

Table 3. Summary of Effect of Placement, Compaction, and New Technologies on In-Place Density

State and section number	MTV	Compaction rollers	Passes (total)	New technology	Density (% of TMD)
1-C	Yes	2 steel wheels	9 static passes each in echelon (18)	None	93.5
1-TS1	Yes	2 steel wheels	2 vibratory and 7 static in each echelon (18)	None	93.2
1-TS2	Yes	2 steel wheels, 1 pneumatic	9 vibratory in each echelon and 9 pneumatic (27)	None	95.4
2-C	No	1 steel wheel	7 vibratory passes (7)	None	91.0
2-TS1	No	1 steel wheel	9 vibratory passes (9)	None	91.8
3A-C	No	2 steel wheels, 2 pneumatic	5 vibratory passes in echelon, 7 pneumatic passes in echelon (24)	MOBA Pave-IR, IC, RDM	92.9
3A-TS1	No	3 steel wheels, 2 pneumatic	5 vibratory passes in echelon, 7 pneumatic passes in echelon, plus 5 vibratory passes (29)	MOBA Pave-IR, IC, RDM	92.9
3A-TS2	No	2 steel wheels, 2 pneumatic	5 vibratory passes in echelon, 7 pneumatic passes in echelon (24)	MOBA Pave-IR, IC, RDM	93.5
3A-TS3	No	3 steel wheels, 2 pneumatic	5 vibratory passes in echelon, 7 pneumatic passes in echelon, plus 5 vibratory passes (29)	MOBA Pave-IR, IC, RDM	94.1
3B-C	No	3 rollers	Not clear which rollers used	MOBA Pave-IR, IC, RDM	93.7
3B-TS1	No	4 rollers	Not clear which rollers used	MOBA Pave-IR, IC, RDM	94.9
3B-TS2	No	3 rollers	Not clear which rollers used	MOBA Pave-IR, IC, RDM	93.6
3B-TS3	No	4 rollers	Not clear which rollers used	MOBA Pave-IR, IC, RDM	94.2
3B-TS4	No	3 rollers	Not clear which rollers used	MOBA Pave-IR, IC, RDM, use of WMA	93.7
3B-TS5	No	3 rollers	Not clear which rollers used	MOBA Pave-IR, IC, RDM, use of WMA	93.8
4-C	Yes	2 steel wheels, 1 pneumatic	5 vibratory passes in echelon, 11 pneumatic passes (21)	None	93.5
4-TS1	Yes	3 steel wheels, 1 pneumatic	5 vibratory passes in echelon, 11 pneumatic passes, 5 vibratory passes (26)	None	95.0
4-TS2	Yes	2 steel wheels, 1 pneumatic	5 vibratory passes in echelon, 11 pneumatic passes (21)	None	94.6
4-TS3	Yes	3 steel wheels, 1 pneumatic	5 vibratory passes in echelon, 11 pneumatic passes, plus 5 vibratory passes (26)	None	95.4
4-TS4	Yes	2 steel wheels, 1 pneumatic	5 vibratory passes in echelon, 11 pneumatic passes (21)	Use of WMA	92.5
4-TS5	Yes	2 steel wheels, 1 pneumatic	5 vibratory passes in echelon, 11 pneumatic . passes (21)	Use of WMA	93.4
4-TS6	Yes	2 steel wheels, 1 pneumatic	5 vibratory passes in echelon, 11 pneumatic passes (21)	Use of WMA	94.0
4-TS7	Yes	2 steel wheels, 1 pneumatic	5 vibratory passes in echelon, 11 pneumatic passes (21)	None	95.2
5-C1	Yes	2 steel wheels, 1 pneumatic	5 vibratory passes in echelon, 5 pneumatic passes (15)	None	92.5
5-TS1	Yes	2 steel wheels	5 oscillatory passes in echelon (10)	None	93.2
5-TS2	Yes	2 steel wheels, 1 pneumatic	5 vibratory passes in echelon, 5 pneumatic passes (15)	None	95.2
6-C	Yes	2 steel wheels	5 vibratory passes and 2 static passes in echelon (14)	Longitudinal joint adhesive	93.3

(continued)

Table 3. (continued)

State and section number	MTV	Compaction rollers	Passes (total)	New technology	Density (% of TMD)
6-TS1	Yes	2 steel wheels	5 vibratory passes and 2 static passes in echelon (14)	Longitudinal joint adhesive	95.4
7-C1	Yes	3 steel wheels	4 vibratory passes and 1 static pass in echelon (15)	WMA	94.4
7-TS1	Yes	3 steel wheels	4 vibratory passes and 1 static pass in echelon (15)	WMA	96.1
8-C1	Yes	1 steel wheel, 1 pneumatic	8 vibratory and 1 static pass, 15 pneumatic passes (24)	None	93.1
8-TS1	Yes	1 steel wheel, 1 pneumatic with increased weight	8 vibratory and 1 static pass, 15 pneumatic passes (24)	None	93.0
9-C	Yes, but not effective owing to mechanical problems	2 steel wheels	3 vibratory passes and 6 static passes, 7 static passes (16)	WMA, MOBA Pave-IR, IC	92.2
9-TS1	Yes, but not effective owing to mechanical problems	2 steel wheels	7 vibratory and 2 static passes, 2 oscillatory passes, and 1 static pass (12)	WMA, MOBA Pave-IR, IC	92.0
9-TS2	Yes, but not effective owing to mechanical problems	2 steel wheels	7 vibratory passes, 2 oscillatory and 3 static passes (12)	WMA, MOBA Pave-IR, IC	92.0
10-C	No	2 steel wheels	9 vibratory passes, 9 vibratory passes (18)	None	95.6
10-TS1	Yes	2 steel wheels	9 vibratory passes, 9 vibratory passes (18)	MOBA Pave-IR, IC, RDM	95.0
10-TS2	Yes	2 steel wheels	9 vibratory passes, 9 vibratory passes (18)	MOBA Pave-IR, IC, RDM	95.7

Note: IC = intelligent compaction; RDM = rolling density meter; TMD = theoretical maximum density; WMA = warm-mix asphalt.

testing be conducted on the asphalt mixture to evaluate its resistance to rutting, cracking, and moisture damage. It is important that the adjusted mix be used only when testing has shown that the performance of the adjusted mix is satisfactory.

Consistency and Uniformity

An advantage of a PWL (percent within limits) specification over the minimum lot average specification is that the consistency, as measured by the standard deviation, is included as part of the specification. The consistency is an important factor. Two of 10 states (states 7 and 8) demonstrated improvements in the standard deviation, and showed that standard deviations below 1.00 were possible, as shown in Table 5. State 8 focused on consistency and lowered the standard deviation sufficiently to effectively raise the lower specification limit by 1.0%.

The New York State Department of Transportation (NYSDOT) was identified as a success story for using the PWL specification. The NYSDOT 50 Series is used on interstates and principal arterials with full or partial control of access. The density is measured with cores. The lower specification and upper specification limits were set at 92.0 and 97.0%, respectively. There is a 5% incentive available on density alone. For 2015, the statewide average density was 94.1%. As observed by NYSDOT, contractors understand that PWL specifications require a focus on consistency in addition to the average density and are focusing on being more consistent. The standard deviation of projects statewide was 0.83.

Incentives can be a valuable part of the specification to gain improvements in in-place density. Seven of 10 states (states 1, 3, 5, 6, 7, 8, and 10) used incentives. All of these, except for state 3, used PWL specifications. Several states noted the importance of the incentive to the success of their improvements in in-place density. Four of the 10 states (states 1, 5, 7, and 9) had a maximum or upper specification limit on density.

Table 4. Summary of the Field Compaction Effort in the Test Sections

State and section number	Total Rollers	Total Passes	Breakdown roller	Field density (% TMD)	Difference from control	Specification	Requirement	Incentive/ disincentive
2-TS1	1	9	No echelon	91.8	+0.8	Minimum sublot (of lab compacted)	96.0	D
9-TS1	2	10	No echelon	92.0	–0.2	Percentage control strip	98.0	D
6-TS1	2	14	Echelon	95.4	+2.1	PWL	91.0	I / D
5-TS2	3	15	Echelon	95.2	+2.7	PWL	92.0	I / D
7-TS1	3	15	Echelon	96.1	+1.7	PWL	92.0	I / D
10-TS2	2	18	No echelon	95.7	+0.1	PWL	92.0	I / D
8-TS1	2	24	No echelon	93.0	+1.0	PWL	91.0	I / D
4-TS3	4	26	Echelon	95.4	+1.9	Minimum lot average	91.5	D
1-TS2	3	27	Echelon/No vibratory	95.4	+1.9	PWL	91.8	I / D
3-TS3	5	29	Echelon	94.1	+1.2	Minimum lot average	92.0	I / D

Note: PWL = percent within limits; TMD = theoretical maximum density.

Table 5. In-Place Density Results for States 7 and 8

State and section number	Average (% TMD)	Standard deviation	Notes
7-C1	94.4	2.7	Density measured using field cores
7-TS1			
Lot 1	96.1	0.6	
Lot 2	96.2	0.8	
Lot 3	96.1	1.0	
8-C1	93.1	1.58	Density measured using the nuclear gauge that has been correlated to cores
8-TS1	93.0	0.67	

Note: TMD = theoretical maximum density.

Best Practices

The number of compaction rollers varied from as few as one roller on one of the field demonstration projects (state 2) and to five compaction rollers on another field demonstration project (state 3). This is a tremendous difference in compaction effort. A summary of some key observations included:

- A summary of the total number of passes on the test section was:
  - Two of 10 states used <12 passes (states 2 and 9);
  - Four of 10 states used 12–20 passes (states 5, 6, 7, and 10); and
  - Four of 10 states used >20 passes (states 1, 3, 4, and 8).
- When vibratory or oscillatory rollers were used, generally all the passes were used the vibratory and/or oscillatory mode. In some cases, there may have been a final one or two passes that were static. However, one state did not use the vibratory mode as much.
  - Two of 10 states used the vibratory mode of the roller in a limited manner in the control section (states 1 and 9).
- A summary of states in which rollers were used in echelon included:
  - Six of 10 states used breakdown rollers in echelon (states 1, 3, 4, 5, 6, and 7); and
  - One of 10 states used intermediate pneumatic rollers in echelon (state 3).
- A summary of states that used pneumatic rollers included:
  - Five of 10 states used pneumatic rollers (states 1, 3, 4, 5, and 8).
  - It should be noted that none of these projects that used pneumatic rollers used polymer-modified asphalt.

The best practices for the number of rollers, number of passes, and number of passes using vibration can be used to improve density.

States 2 and 9 clearly had the lowest density values in their test section compared to the other states. State 2 used one roller for compaction but was still able to make an improvement in density by using just two more passes. This showed that the amount of rolling significantly affected the density. State 9 used one vibratory roller and one oscillating roller but with very few passes. State 9 was the only state not able to make an improvement in in-place density in their test section. Based on the field report, they used fewer passes in their test section than in the control section.

Table 6. State 4 Cost–Benefit Analysis for 1 Ton of Mixture

Item	Benefit	Cost	Increase in percentage density
Life cycle cost	$6		
Additional roller		≤$1	+1.9
Engineered mix design adjustment		≤$2	
WMA additive		≤$1	–
Smaller NMAS		≈$2	+1.7

Note: NMAS = nominal maximum aggregate size; WMA = warm-mix asphalt.

An additional roller was helpful in increasing density in the test section. Three of 10 states used an additional roller: states 1, 3, and 4. These states were all successful at obtaining higher density. State 1 obtained higher density with the addition of a pneumatic roller. Additional density could not be achieved with only an increase in passes with the steel-wheeled rollers.

State 4 conducted a cost–benefit analysis for using an additional roller. A summary is shown in Table 6. Based on a previous analysis (*2*), a 1% increase in in-place density could result in a 10% improvement in life cycle cost, conservatively. For an asphalt mixture that costs $60 per ton, a 10% improvement in life cycle cost was approximately $6, as shown in Table 6. State 4 also estimated the costs of the adjustments made to increase density as part of the field demonstration project for comparison. Based on the results of this cost–benefit analysis, State 4 believes these adjustments were cost-effective. Some of these adjustments when combined can potentially increase the in-place density by more than 1%, resulting in better cost savings.

It should be noted that there are many best practices to achieve higher density other than adding a roller, which could include: roller settings, vibration versus speed, mat temperature, vibrating screed, paver speed, and so forth. These best practices may be less costly than an additional roller. Many of these practices were outside the scope of each of the SHA's field demonstration project.

During placement and compaction, there were a surprising number of issues with equipment operation. Five of the 10 states had equipment-related issues (states 1, 4, 8, 9, and 10) during construction of their sections. In each of these five field demonstration projects, the equipment issue was an impediment to achieving higher density. An agency may need to require a quality-control plan to make sure compaction equipment is working properly before paving.

New Technologies (Primarily States 3 and 10)

Several states evaluated new technologies to help ensure good compaction. The technologies used included warm-mix asphalt (WMA), MOBA Pave-IR System, rolling density meter (RDM), and intelligent compaction (IC). The number of states using each of the technologies was as follows:

- WMA was used by six of the 10 states (states 3, 4, 5, 7, 9, and 10).
- MOBA Pave-IR System was used by three of the 10 states (states 3, 9, and 10).
- IC was used by three of the 10 states (states 3, 9, and 10).
- RDM was used by two of the 10 states (states 3 and 10).

These new technologies showed some promise. However, although these new technologies were used and data were collected, very little analysis of data was made in most cases, particularly in real time. Most of these technologies have the potential to improve the quality of large projects but were not very effective when used in small sections as used in this demonstration project. These technologies generally provided information that would have been useful in adjusting as work progresses so would be most useful for larger projects. The application of these new technologies was evaluated, but they had little benefit in placement and compaction of these small sections.

A potential benefit of some of these new technologies is that there will be more test results and better quantification of in-place materials. There can be challenges with new technologies, such as issues obtaining real-time data to make project adjustments. Also, the use of some of the results for acceptance purposes has not been fully demonstrated to be accurate. Although these new technologies may be a good quality-control tool, care should be exercised for acceptance.

Summary

There are significant differences in specifications, in amount of rolling effort, and in method of specifying in-place density between states, but the field demonstration projects show that density can be improved. Eight of the 10 states improved densities by at least 1% on their field demonstration projects. Seven of the 10 states improved densities to a minimum of 94.0% of the TMD. A summary of the methods used to obtain increased density seemed to fall into one of the following five categories.

1. Improving the agency's specification by including or increasing incentives and examining the minimum percentage density requirements. There was a significant difference in the number of rollers used for compaction between states. Some states used as little as one compaction roller while others used as many as four or five compaction rollers. The number of passes for each roller used varied considerably among states. There is a strong association between the rolling effort and the agency's requirements. Some states

could obtain high density in the range of 95.0–96.0% of TMD, while other states only obtained density in the range of 90–91%. As expected, using fewer rollers and fewer vibratory passes generally resulted in lower in-place density, and using more rollers resulted in higher in-place density.

2. Some states obtain higher density by increasing the optimum asphalt binder content with engineering adjustments. Making engineering adjustments to the asphalt mixture design to obtain slightly higher optimum asphalt content was successful at achieving higher in-place densities. These adjustments included either reducing the number of gyrations during mix design or designing at lower air voids. It is important that a satisfactory mix be designed and produced to ensure good performance and that this mix be compacted to adequate density in the field. As a word of caution, adding additional asphalt solely for compaction changes the mixture properties and this adjusted mix should only be used if laboratory test results have shown that this adjusted mix is satisfactory.
3. Consistency is one of the most important factors in improving in-place density. Consistency can be generally defined as consistency in temperatures, paver speeds, roller patterns, and all the other factors that impact density and standard deviation of density measurements. Improving consistency as measured by the standard deviation was accomplished by two of the states. Information from states 7 and 8 demonstrated that a standard deviation below 1.00 was possible and could be achieved routinely.
4. Following best practices is important. There was a lot of attention on the construction of the control and test sections. Since these pavement sections were built for the field demonstration projects, there was more attention to best practices than there would normally have been. It should be noted that there are many best practices other than those used for these field demonstration projects. Although the states tried many things, there were likely many other best practices that could have made a difference. Best practices other than adding another roller could include: roller settings, vibration versus speed, mat temperature, vibrating screed, paver speed, and so forth. These best practices may even be less costly than using an additional roller.
5. Using new technologies was helpful. These technologies used included WMA, MOBA Pave-IR System, RDM, and IC. These new technologies showed some promise.

Not all of these methods may work for every state. However, the list can serve as a toolbox or checklist of considerations to identify areas for improvement.

Acknowledgments

The authors wish to acknowledge the funding by the FHWA. The authors would like to acknowledge the many parties who helped make this demonstration possible. The authors thank the Asphalt Institute for delivery of the 10 workshops before the field construction of the demonstration projects. The authors would like to thank the key contacts for coordinating the compaction workshop and field demonstration project in each of the states. Many people were involved with this process. The key contacts from the SHAs and FHWA Division Offices included Alaska, District of Columbia, Florida, Indiana, Minnesota, Oklahoma, Pennsylvania, Virginia, Washington State, and Wisconsin.

References

1. Asphalt Institute. *The Asphalt Handbook*. 7th ed. Asphalt Institute, College Park, MD, 2007, pp. 788.
2. Tran, N., P. Turner, and J. Shambley. *Enhanced Compaction to Improve Durability and Extend Pavement Service Life: A Literature Review*. NCAT report no. 16-02. National Center for Asphalt Technology, Auburn, AL, 2016, pp. 25.
3. Epps, J., C. Monismith, J. Deacon, J. Harvey, and R. Leahy. *NCHRP Report 455: Recommended Performance-Related Specification for Hot-Mix Asphalt Construction: Results of the Westrack Project*. HRB, National Research Council, Washington, D.C., 2002.
4. Timm, D., R. West, A. Priest, B. Powell, I. Selvaraj, J. Zhang, and R. Brown. *Phase II NCAT Test Track Results*. NCAT report no. 06-05. National Center for Asphalt Technology, Auburn, AL, 2006, pp. 83.
5. Brown, E. R., M. Hainin, A. Cooley, and G. Hurley. *NCHRP Report 531: Relationship of Air Voids, Lift Thickness, and Permeability in Hot Mix Asphalt Pavements*. HRB, National Research Council, Washington, D.C., 2004.
6. Hekmatfar, A., R. McDaniel, A. Shah, and J. Haddock. *Optimizing Laboratory Mixture Design as It Relates to Field Compaction in Order to Improve Hot-Mix Asphalt Durability*. Joint Transportation Research Program Technical Report SPR-3624. Purdue University, West Lafayette, IN, 2013, pp. 51.
7. Federal Highway Administration. *Superpave Mix Design and Gyratory Compaction Levels*. Tech Brief FHWA-HIF-11-03. Office of Pavement Technology, Federal Highway Administration, Washington, D.C., 2010, pp. 10.
8. Beainy, F, D. Singh, S. Cummuri, and M. Zaman. Laboratory and Field Study of Compaction Quality of an Asphalt Pavement. *International Journal of Pavement Research and Technology*, Vol. 7, No. 5, 2014, pp. 317–323.
9. Scherocman, J. Compaction of Stiff and Tender Asphalt Concrete Mixes. *Factors Affecting Compaction of Asphalt Pavements*. Transportation Research Circular E-C105. Transportation Research Board, Washington, D.C., 2006, pp. 69–83.
10. National Asphalt Pavement Association. *Balancing Production Rates in Hot Mix Asphalt Operations*. Information Series 120. National Asphalt Pavement Association, Lanham, MD, 1996, pp. 32.
11. Timm, D. MultiCool Software. http://www.eng.auburn.edu/users/timmdav/MultiCool/FinalRelease/Main.html#Disclaim. Accessed February 20, 2017.

12. Federal Highway Administration. *A Review of Aggregate and Asphalt Mixture Specific Gravity Measurements and Their Impacts on Asphalt Mix Design Properties and Mix Acceptance*. Tech Brief FHWA-HIF-11-033. Office of Pavement Technology, Federal Highway Administration, Washington, D.C., 2010, pp. 39.
13. Santucci, L. *The Role of Compaction in the Fatigue Resistance of Asphalt Pavements*. Institute of Transportation Studies, Technology Transfer Program, University of California at Berkeley, Berkeley, CA, 1998.
14. Nodes, J. Impact of Incentives on In-Place Air Voids. *Factors Affecting Compaction of Asphalt Pavements*. Transportation Research Circular E-C105. Transportation Research Board, Washington, D.C., 2006, pp. 163–168.
15. Aschenbrener, T., E. R. Brown, N. Tran, and P. Blankenship. *Demonstration Project for Enhanced Durability of Asphalt Pavements through Increased In-Place Pavement Density*. NCAT report no. 17-05. National Center for Asphalt Technology, Auburn, AL, 2017, pp. 86.
16. Prowell, B., G. Hurley, and B. Frank. *Warm Mix Asphalt: Best Practices*. 3rd ed. National Asphalt Pavement Association, Lanham, MD, 2012.
17. Chang, G., Q. Xu, J. Rutledge, B. Horan, L. Michael, D. White, and P. Vennapusa. *Accelerated Implementation of Intelligent Compaction for Embankment Subgrade Soils, Aggregate Base, and Asphalt Pavement Materials*. Final report. FHWAIF-12–002. Federal Highway Administration, Washington, D.C., 2011.
18. Chang, G., Q. Xu, J. Rutledge, and S. Gerber. *A Study on Intelligent Compaction and In-Place Density*. FHWA-HIF-14-017. National Technical Information Service, Washington, D.C., 2014.
19. Willoughby, K., J. Mahoney, L. Pierce, J. Uhlmeyer, K. Anderson, S. Read, S. Muench, T. Thompson, and R. Moore. *Construction-Related Asphalt Concrete Pavement Temperature Differentials and the Corresponding Density Differentials*. Report Number WA-RD 476.1. Washington State Transportation Center, University of Washington, Washington, D.C., 2001.
20. Benson, J., and J. Scherocman. Construction of Durable Longitudinal Joints. *Transportation Research Circular E-C105: Factors Affecting Compaction of Asphalt Pavements*. TRB, National Research Council, Washington, D.C., 2006.
21. Asphalt Institute. *State Highway Agency (SHA) Density Specification Data Mining*. FHWA Co-op. Task 2.15 State Density Maps. Asphalt Institute, College Park, MD, 2016. http://www.asphaltpavement.org/index.php?option=com_content&view=article&id=1049&Itemid=100334. Accessed February 20, 2017.
22. Federal Highway Administration. *Tack Coat Best Practices*. Tech Brief FHWA-HIF-16-017. Office of Pavement Technology, Federal Highway Administration, Washington, D.C., 2016, pp. 17.

The Standing Committee on Asphalt Pavement Construction and Rehabilitation (AFH60) peer-reviewed this paper (18-00583).

Article

TRR
JOURNAL OF THE TRANSPORTATION RESEARCH BOARD

Transportation Research Record
2018, Vol. 2672(26) 68–76
© National Academy of Sciences:
Transportation Research Board 2018
Article reuse guidelines:
sagepub.com/journals-permissions
DOI: 10.1177/0361198118773888
journals.sagepub.com/home/trr
SAGE

Design, Fabrication, Construction, and Cost Assessment of Virginia's First Corrosion Resistant ASTM A1010 Plate Girder Bridge

Stephen R. Sharp[1], Jason Provines[1], Audrey K. Moruza[1], William F. Via , Jr.[2], and Keith N. Harrop[2]

Abstract

This study documents the construction of the Waynesboro Route 340 Bridge with low-maintenance corrosion resistant steel girders, cross-frames, and stainless steel bolts. This includes documenting the fabrication of the ASTM A1010 steel plate, girder and secondary members, erecting the girders, and completion of the construction process. The types of stainless steel materials used to fabricate the girders are reported with a comparison to traditional materials. Information on the material properties was documented. Cost analysis was performed and guidance was provided to the Virginia Department of Transportation (VDOT) about how this type of plate girder will benefit the DOT. Overall, the design, fabrication, and construction of the A1010 steel girder Route 340 bridge was a success. This is the first bridge in the United States to use A1010 haunched girders, completely stainless steel bolted splices, and A1010 cross frames. To that end, it was concluded that a conventional steel bridge fabricator can successfully fabricate an A1010 steel bridge. Through detailed testing of the stainless steel bolts, it was also shown that acceptable utilization of stainless steel structural connections can be achieved. VDOT was also able to successfully demonstrate fabrication of A1010 steel cross frames utilizing bent plate shapes. It is recommended to continue to evaluate improved welding processes for the A1010 base metal and to consider the use of A1010 as a repair material for in-service structures located in corrosive macro-environments, such as near saltwater or industrial sites, or those which contain aggressive micro-environments such as steel beam ends under joints.

Corrosion of carbon steel structures is a concern for the Virginia Department of Transportation (VDOT). Along both the coast and inland, steel can be exposed to chloride-containing salts, which further promote corrosion of the steel. Along the coast, structures are subjected to both brackish waters and seawater. Moving inland, structures are exposed to deicing salts and brine that are used to minimize travel disruptions and unsafe road conditions during snow and ice events. Furthermore, the use of uncoated weathering steel can be limited by the proximity of this steel to water or industrial environments where there are concerns over the proper performance of the protective patina on the steel surface. Given the difficulty of maintaining applied protective barriers on the steel and the continuing use of de-icing salt treatment on roads, another option is to use materials that exhibit greater resistance to chloride-induced corrosion, thus providing longer low-maintenance service lives.

Transportation system maintenance cost increases in recent years have meant that VDOT and other Departments of Transport (DOTs) have started to investigate and use materials with higher initial cost but provide increased durability over the structure's service life. For example, VDOT has used stainless steel in bridge decks and prestressed concrete piles, and carbon fiber composite cable has been used in prestressed concrete piles and beams. These materials always have higher initial construction costs but have been approved on the expectation of extended service lives and greatly reduced maintenance expenditures over the life of the structure. Also, when measuring the cost premium of such materials against the total cost of the construction project, some of these premiums are relatively small. Therefore, life-cycle cost analysis of a premium material option should be performed to guide conclusions about overall cost-effectiveness, that is, whether avoided future maintenance (and/or

[1]Virginia Transportation Research Council, Charlottesville, VA
[2]Virginia Department of Transportation, Richmond, VA

Corresponding Author:
Address correspondence to Stephen R. Sharp:
Stephen.Sharp@VDOT.Virginia.gov

replacement) and traffic control operations over the project's service life are expected to offset the higher initial cost of a premium construction material.

Several years ago, VDOT became aware of one innovative material for mitigation of both corrosion and maintenance cost of steel plate girders: ASTM A1010 steel. The material, which was recently added to ASTM A709 as Grade 50CR, is currently offered as a plate product and has shown great potential for uncoated corrosion resistant plate girders. Several years ago, the Federal Highway Administration (FHWA) performed a study titled, "Improved Corrosion-Resistant Steel for Highway Bridge Construction" (*1*). In this study, the corrosion resistances of several steel plate products of different compositions were compared. The results demonstrated superior cost-effectiveness of A1010 over the service life of the project, subject to the assumptions of the economic analysis (*1*).

ASTM A1010 steel plate has been used to build structures in California, Pennsylvania, Oregon, Iowa, and Canada (*2*, *3*). The first A1010 steel bridge was constructed in California as a multi-cell box girder bridge. For this bridge A1010 was selected because of the close distance between the girders and the water. In Pennsylvania, A1010 was used to fabricate plate girders that were used for a bridge at the steel producer's Coatesville facility. Since then Oregon DOT has constructed two highway bridges, and both Iowa DOT and the Ministry of Transportation in Ontario, Canada have built a bridge using A1010 steel.

The VDOT Staunton District decided to replace the Route 340 bridge, which is located within a city park and leads into the City of Waynesboro's Downtown Historic District. The decision to replace versus rehabilitate this bridge was because of several facts. The old structure was considered structurally deficient, with it being posted for a 15 ton weight restriction. The existing abutments and piers had deteriorated, receiving a 4 out of 9 rating on the last inspection report. Finally, the existing superstructure included a non-redundant fracture critical pin and hanger system that would require substantial effort and cost to eliminate if the bridge were rehabilitated.

The Staunton District decided instead to incorporate ASTM A1010 steel for the primary and secondary members as part of the Route 340 replacement project. The rationale for selecting low-maintenance ASTM A1010, as well as stainless steel hardware, to connect the structural members for this seven million dollar project was because of a combination of FHWA guidelines on the use of uncoated weathering steel and the desires of the local governing officials (*4*). The primary concerns that governed this decision are as follows:

- Proximity to water: the bottom flange elevation of the proposed Route 340 bridge will be roughly 9' above the South River.
- Flood events exceeding a 10 year storm will reach the bottom flange elevation of the proposed bridge.
- A chemical plant is upstream (within view from the bridge site).
- Aesthetics: the bridge is situated above a city park that leads to the historic district.

Research into ASTM A1010 is in line with the VDOT transition from the use of coated steels in reinforced concrete deck construction to uncoated, corrosion-resistant alloyed steel to reduce maintenance costs associated with corrosion in bridge decks. The Route 340 bridge is an example of the transition of the VDOT, as the bridge deck supported by the A1010 girders was constructed using corrosion resistant type 2304 stainless steel reinforcing bars for the deck reinforcement.

This transition to using corrosion resistant materials demonstrates the agency's perspective that the selection of structural steel should no longer be based exclusively on initial costs; the decision should be based on the initial construction costs *plus* the expected future costs associated with steel maintenance and repair operations (*5*). Ideally, the decision should be supported by means of life-cycle cost analysis, and an economic analysis framework, which is particularly useful when there is a possibility that a slight increase in material costs at the construction phase could significantly reduce long-term maintenance costs (*5*).

Purpose and Scope

The purpose of this project was to validate that ASTM A1010 steel plate girders can be technically and cost-effectively employed in Virginia for bridge structures at high risk for corrosion. Validation involved monitoring and evaluation of the handling of A1010 steel, acceptance by steel fabricators, comparison of A1010 fabrication issues with traditional methods and materials, monitoring the behavior of the girders during erection and after placement, and bolting operations. A past VTRC study on the use of stainless steel bolts for A1010 steel girders was also conducted to determine the best-suited bolt material for the Route 340 bridge as a function of the physical performance and cost and was published in the Transportation Research Record (*6*). As ASTM A709 Gr. 50W was not well-suited for this structure, an in-depth cost analysis was desired for using ASTM A1010 steel plate girders.

Research Materials

The steel selected for the plate girders of the Route 340 bridge was specified to meet the requirements of ASTM A1010. This standard stipulates that the chemical and mechanical properties required for martensitic stainless steels that will be used in structural applications and limits the maximum plate thickness to 2 inches (*7*). For the bridge, A1010 Grade 50 was specified, which has a yield strength of 50 ksi and an ultimate tensile strength of 70 ksi.

For welding, the steel producer recommended that an austenitic stainless steel filler metal be used (3). One recommended metal that has been used successfully is Type 309L, where the "L" signifies low carbon content. This type of filler is recommended for welding either carbon steel to stainless steel, or stainless steel to stainless steel and exhibits favorable resistance to intergranular attack because of the low carbon content of the electrode (8).

The fasteners selected were ASTM A193 B8 bolts and ASTM A194 B8 nuts. These are austenitic stainless steel materials and are more commonly known by their SAE designation of Type 304. These bolts are required to exhibit a minimum tensile strength of 100 ksi and provide a minimum clamping load of 30 kips per bolt at the splice connection.

Results and Discussion

Route 340 Bridge Design

In many ways, the design of the Route 340 bridge was similar to that of a traditional carbon steel girder bridge; both A1010 and A709 Grade 50, a typical bridge steel, have yield strengths of 50 ksi and ultimate strengths of 70 ksi. However, there were some differences because of the new materials used. The previous structure over the South River was a 4-span steel girder bridge, with the piers oriented perpendicularly to the roadway. As the piers were not parallel to the river, they had disrupted the natural flow of the river, creating an island downstream of one of the piers. The soil boring logs at the bridge location showed that the depth of bedrock was much deeper than other nearby areas. This meant that replacing piers of the previous structure with new ones that meet today's vehicle loads and design guidelines would be expensive. Owing to the disruption of the river flow and the cost of replacing the piers, the decision was made to design the new structure using a single pier, oriented parallel to the river; this changed the span layout from the previous 4-span structure to a new 2-span bridge, and increased the span lengths to approximately 110 ft.

The hydraulic study revealed that the elevation of the finished grade of the deck elevation along the proposed bridge centerline needed to remain at the same elevation as the previous structure to prevent an increase in the water level during a flood event. As the floodplain in the area surrounding the bridge is relatively wide, a small increase in the water level could cause flooding to many businesses located in downtown Waynesboro. The bottom flange elevation of the structure was also limited by a walkway under the west side of the bridge, which requires 8 ft. clearance. In addition, because of the close proximity of the surrounding business entrances, it was not feasible to raise the top of the grade elevation given the approach roadway geometrics. The combination of the high and bottom flange elevation limitations created girder elevation and depth restrictions.

The new span layout and depth limitations of the new bridge were not at all related to the new material used on the bridge; however, the ASTM A1010 specification limits the steel to a maximum thickness of 2 in. The combination of the span layout, depth limitation, and maximum thickness of the material led to the selection of a haunched steel girder design, particularly to control stresses in the negative moment region over the pier. Haunched girders are not uncommon but do require the designer to consider an alternative to a more typical constant depth plate girder. Some benefits of using the haunched girder design are that the girders are typically lighter weight than constant depth girders, and are seen as aesthetically pleasing, which was of interest as the bridge is located in a city park.

Another difference in the design of the bridge occurred in the design of the secondary members. Other materials were considered for the cross frames as other agencies had used weathering steel or galvanized steel with A1010 girders. If one of these options were selected, it is possible that the cross frames could require maintenance or even replacement at some point during the life of the structure. VDOT elected to use A1010 steel for the cross frames so that all of the steel plate used for the structure would have the same design life when considering the corrosion resistance. As A1010 is currently only available in plate form, and not in rolled shapes, VDOT allowed the fabricator to choose whether the cross frames would be bent into shapes or welded into built-up members. The fabricator elected to cold bend the material into angle and channel shapes, which were eventually welded together to form the cross frames.

The surface finish of all of the steel on the bridge, including the girders and secondary members, was also different because of the use of A1010. For a typical carbon steel bridge, after the bridge elements have been fabricated, they can either be left in the unblasted condition with the mill scale present on the steel, or they can be blast cleaned. If they are blast cleaned, it is typically done using a steel shot grit. At the time of the design, it was recommended that A1010 be blast cleaned to prevent the mill scale from creating a nonuniform weathering pattern. Sand, ceramic, or non-iron blasting media were recommended as the use of steel shot could create a speckled rust pattern if improperly done (3). For this reason, VDOT specified a non-metallic blast media in hopes of creating a uniform, rustic patina so the bridge would complement the surrounding park-like setting. As such, the fabricator elected to use a garnet media for the blast cleaning.

Another significant difference between the design of the Route 340 bridge and a traditional steel bridge was the bolted splice design. The other A1010 bridges in the United States had been constructed using either weathering steel or galvanized steel bolts. To provide a truly low maintenance structure, VDOT opted to use stainless steel bolt, nut, and washer fastener assemblies. After conducting an experimental and cost analysis on a variety of grades of stainless steel, ASTM

A193 Grade B8 steel bolts, ASTM A194 steel nuts, and Grade 303 steel washers were selected. An in depth discussion of the rationale for selecting the Grade B8 bolts and accompanying fastener assemblies can be found in the Transportation Research Record (*6*).

The stainless steel Grade B8 bolts have an ultimate strength of 100 ksi, whereas ASTM F3125 Grade A325 bolts, those typically used in steel bridges, have an ultimate strength of 120 ksi. An assumption also had to be made about the surface condition factor of the faying surfaces. Although neither A1010 steel nor garnet blasted faying surfaces are included in the AASHTO LRFD specification, it was assumed the behavior of this type of connection would be similar to A709 steel with a blast cleaned surface. This led to using a Class B surface condition in the splice design process. All of the bolts used on the bridge were 7/8 in. diameter, as this is the VDOT standard bolt size. The lesser ultimate strength meant that the minimum clamping force on each stainless steel bolt was reduced from 39 kips (standard Grade A325 bolt) to 30 kips.

Fabrication using A1010 Plate

VDOT officials developed a set of special provisions for the contractor to provide additional specifications related to the unique nature of this bridge. The special provision provided the fabricator with requirements for the fabrication operations. Cutting, machining, bending, and welding procedures were monitored to determine the effectiveness of the special provision and ascertain areas of improvement. Monitoring was done using visual analysis on the as-received A1010 plates and on the final product. Visual analysis and nondestructive testing were also performed on the A1010 plates during fabrication.

A1010 steel slabs were made at a PA facility of the steel producer, Coatesville. For rolling thick plates, the slabs remained at this facility, but for rolling thin plates, slabs were shipped to a second rolling facility that specializes in rolling thinner plates. Fabrication of the A1010 into structural members was then done by a bridge fabricator at two of their facilities, one in Lancaster, PA and the other in Williamsport, PA.

Mill Markings and Cutting

Before beginning any fabrication, mill markings (Figure 1*a*) were checked. It was quickly evident that while most of the information was provided on the steel plates, the steel grade was not. This is important as ASTM A1010 covers both Grade 40 and Grade 50 steels. This observation was reported to the steel producer to ensure this it is corrected for future use.

The A1010 steel plates were cut (Figure 1*b*) in Lancaster, PA, where secondary member components remained to be fabricated; primary member components were shipped to Williamsport, PA for fabrication. The plates were cut using a plasma cutter as A1010 steel cannot be cut with the traditional oxy-fuel technique. Care was required during cutting because of the higher chromium content of the steel. To ensure that exposure to hexavalent chromium in fumes did not exceed Occupational Safety and Health Administration (OSHA) or fabricator limits, the plasma cutter was equipped with a downdraft table, workers wore masks as needed, and the air-quality was monitored.

Machining

As discussed in other studies, the machining (Figure 1*c*) of ASTM A1010 was found to be similar to other Grade 50 bridge steels (*2*, *9*). VDOT contract documents originally required the fabricator to dedicate new tools to be used exclusively on A1010 during the entire machining process, with tools used on carbon steel being allowed only if approved by VDOT. The intent of using dedicated tools was specified to prevent the introduction of carbon into areas which were to be welded. During the fabrication process, the language in subsequent contract documents were revised so that dedicated tools were not required to machine areas of the steel that would not be welded later, such as drilling bolt holes.

Bending

Initially, it was unclear if all of the bending for the secondary members would be done by the fabricator or if a portion of it would be done by a subcontractor. Ultimately only the waterline supports and diaphragms were bent by a subcontractor, though each were sent for different reasons; the waterline supports were sent to a subcontractor because of the precision of the bend required and the diaphragms were sent because of the thickness of the plate material. All other bending, including the steel for the cross frames (Figure 1*d*), was bent in-house by the fabricator without issue.

Welding

An area in the facility was dedicated to the welding of A1010 plates at the Lancaster and Williamsport facilities, and dedicated tools were required during welding (Figure 1*e*). Welding of the primary members was performed in Williamsport and the secondary members were welded in Lancaster. During welding, air-quality was monitored and the equipment was fitted with fume collection devices as needed to prevent harmful exposure to fumes.

During the design and preconstruction stage, the intention of the VDOT was to closely follow the work of the Oregon DOT with the welding of the A1010, therefore only Lincoln material was specified (*2*). However, difficulties in procuring these materials resulted in adjustments being made during fabrication. The submerged arc welding (SAW) process (Figure 1*f*) used for the complete joint penetration (CJP)

Figure 1. ASTM A1010 (*a*) mill marking, (*b*) ½-in thick steel being cut using a plasma torch, (*c*) girder having holes drilled using conventional equipment, (*d*) typical approach for bending a plate, (e) dedicated tools are used during fabrication, and (*f*) cut plates are butt welded to create the flanges.

welds were conducted with LincolnWeld 309/309L filler metal and LincolnWeld 880M flux. The flux-cored arc welding (FCAW) process used to tack the web and flange also used the same filler metal. For the fillet welding of the web to flange, the SAW process was used again, with LincolnWeld 309/309L or SelectArc 309L-C filler metal and LincolnWeld 880M flux. Initially the web-to-flange fillet welds were welded in a 2F position (where the web plate is vertical and the flange plate is horizontal), but it was determined that the 1F position (where the web and flange plates are rotated at 45 degrees, so that the top of the weld metal is flat when placed) provided more favorable results. The SAW process was used to weld cross frames before bundling for shipment to Williamsport to be garnet-blasted.

Nondestructive Testing

Several different nondestructive testing (NDT) techniques were used to inspect the welds: dye penetrant (PT), magnetic particle testing (MT), ultrasonic testing (UT), and X-ray testing (RT). DT testing provided successful results for the fillet welds as expected, but the dye residue was difficult to clean

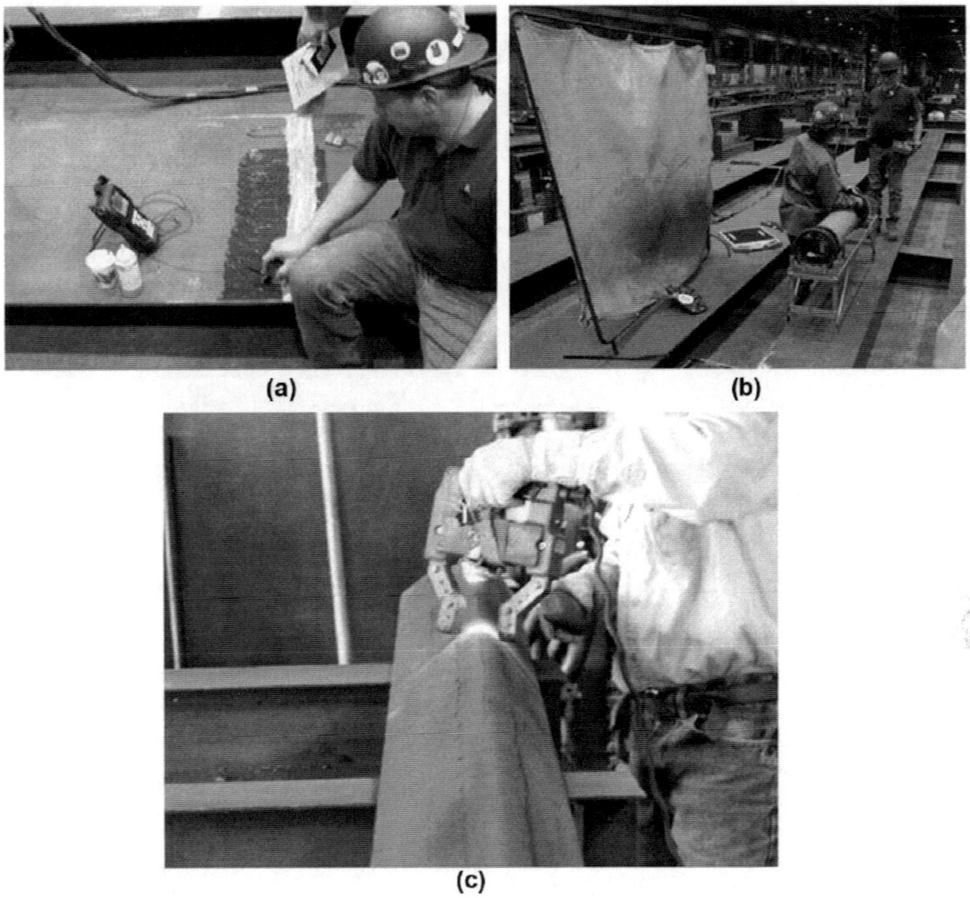

Figure 2. NDT on fabricated ASTM A1010 steel showing (*a*) UT on butt weld, (*b*) RT on butt weld, and (*c*) MT after bending plate.

from the welds and it continued to seep out long after testing took place. MT worked well for evaluating the A1010 plate, especially after bending, but it could not be used to inspect filler metal in the CJP welds because the filler is non-magnetic. UT provided reasonable results, but they were sensitive to the difference in density between the A1010 and filler metal. RT was performed using both film and digital x-ray. Challenges occurred with digital X-ray which might have been because of the difference in A1010 and filler metal as A1010 is a martensitic and ferritic steel and the filler metal is an austenitic steel. This would mean that there are some differences in the lattice structure in the weld region. Better RT results were obtained using film, rather than digital X-ray because of the increased sensitivity associated with using digital X-ray.

Although the NDT techniques proved somewhat challenging because both the base and filler metals are new to steel bridge fabrication, discontinuities were successfully detected on both groove and fillet welds during the fabrication process. All detected discontinuities were repaired and then were retested with passing NDT results. As the NDT techniques were able to find discontinuities and then show passing test results on the repairs, VDOT was confident that traditional NDT techniques were producing accurate test results. The NDT results and the successful procedure qualification record (PQR) test results provided VDOT with confidence that good quality welds were produced on the bridge members. Various NDT techniques are shown in Figure 2.

Erection of the A1010 Steel Plate Girders

As the steel was being erected, observation and documentation (including discussions with field inspectors and the girder erection subcontractor) were performed to determine differences between A1010 and a typical carbon steel. Shear studs can be shop-welded for any bridge project, but VDOT elected for this option for the Route 340 bridge because if, for some reason, problems occurred with the welding of the carbon steel shear studs to the A1010 steel, repairs could be made much easier and the environment is much better controlled in a shop rather than a field environment. The shear studs were successfully welded in the fabrication shop to the A1010 girders without any major challenges.

Figure 3. Erection of Route 340 Bridge A1010 Girders: (*a*) installation of the stainless steel bolts with shear studs welded to girders and (*b*) various stages of the patina on the steel girders.

Other than the carbon steel shear studs, the remaining steel used during the erection of the bridge was alloyed, corrosion-resistant steel. The elastomeric expansion abutment bearings contained A1010 sole plates. Both abutments were semi-integral so no joints were present on the bridge. The pier bearing was a fixed low profile bearing which contained A1010 sole and masonry plates. As the bridge was on a skew, an internal thrust block was incorporated into the design to resist rotational force. The rub plate on this thrust block was made of ASTM A276 Type 316 stainless steel and was anchored onto the concrete thrust block using the same carbon steel shear studs that were used across the tops of the girders.

Another one of the differences between the design of the Route 340 bridge and a traditional steel bridge was the presence of stainless steel bolts. Installation of these bolts required a modified tightening procedure because of their reduced strength when compared with traditional Grade A325 bolts. VDOT officials met with the erection subcontractor before erection of the bridge to provide them with the information necessary to achieve full pretension in the bolts. Once it was time to tighten the stainless steel bolts on the bridge, the subcontractor was able to successfully install and pretension all bolts with negligible difficulty.

An unexpected observation came when the girders were shipped to the jobsite over three separate days. The first shipment of girders was transported from Williamsport, PA to the bridge site in Waynesboro, VA during a clear, sunny day. When the girders arrived at the jobsite, they were a distinct gray color. The second shipment of girders was transported through snow showers and de-icing salt that had been applied to the roadway along the delivery route. When the girders arrived at the site, the patina had already started to form, causing the girders to appear mostly brown. The third and final shipment of girders was transported through rain; once they arrived at the site, they appeared mostly gray with some indications of the brown patina beginning to form. As the only difference between the girders was the environment in which they were delivered, it is believed that de-icing salt was kicked up from the roadway and accelerated the early formation of the expected patina on the girders. There were discussions about possibly accelerating the formation of the patina on the first and third shipment of girders to match the second to provide a uniform, aesthetically pleasing appearance, but ultimately, the decision was made to allow the patina to form naturally on the girder over time. In the future, this aesthetic issue could be addressed in the special provision by changing the blasting technique or by requiring the girders be protected from the weather during shipment and storage.

Figure 3 shows a photo of the installation of the stainless steel bolted splice and the various stages of the patina observed during erection.

Figure 4 shows a photo of the bridge after construction was completed in June 2017.

Documenting and Assessing Costs of Fabrication and Construction of A1010 Steel Girders for the Route 340 Bridge

The contract for the Route 340 bridge replacement was awarded on the basis of Cost + Time bids that consist of two parts: a dollar bid for contract items (Part A) and a dollar bid for road user cost caused by road closure during critical construction (Part B), as defined in the contract. Part B is the product of a bid "Time" to a completion benchmark and a predetermined road user cost per day. In the awarded contract the bid "Time" to completion of critical construction becomes the basis for the application of incentives to complete the work early, penalties for delayed performance, and liquidated damages.

In the Route 340 project, the awarded contract offered a bid of \$7,024,800.74 for Part A and \$1,383,000 (461 days) for Part B, for a total bid of \$8,408,000; the competitor

Figure 4. Photograph of the completed Route 340 bridge in Waynesboro, VA.

Table 1. Incremental Contributions to Unit Cost of ASTM A1010 Girders by Construction Phase

VDOT	RT 340	2016-17	Bid for Item 61800 in contract	Sub payments	"Value added"	Marginal cost/lb
Prime contractor			$ 2,170,000		$ 323,272	$ 0.89
Erection subcontractor				$ 1,846,728	$ 121,875	$ 0.33
Fabrication subcontractor				$ 1,724,853	$ 1,252,403	$ 3.44
Material subcontractor				$ 472,450	$ 472,450	$ 1.30
				Item 61800	$ 2,170,000	$ 5.95

offered a bid of $7,661,781.90 for Part A and $1,455,000 (485 days) for Part B, for a total of $9,117,000. As with ordinary contracts, the lowest total bid was awarded the contract. After issue of the Notice to Proceed on February 26, 2016, the substantial completion deadline was determined to be July 14, 2017 with final completion including landscaping by December 15, 2017.

By July 2017 it was possible to attribute actual costs to project phases by the prime contractor and subcontractor as shown in Table 1, in which the VDOT item code 61800 in the larger bridge contract represents the line item cost of ASTM A1010 girders.

Of the marginal or incremental contributions to the total unit cost of the A1010 girders of $5.95/lb, the cost of fabrication clearly dominates other unit costs of the A1010 girder production. VDOT had anticipated several fabrication-related cost concentrations, including the following:

- Haunch girders, which could potentially cause 8–15% web waste.
- A1010 cross frames; uncertainty about bending and welding the material.
- Stainless steel fasteners; never used with A1010 before this bridge.
- Blast-cleaning of the steel using hand-blasted garnet medium rather than conventional machine-blasted steel shot.
- Control of hexavalent chromium exposure during cutting and welding of A1010 plates.
- Protection of weld area from carbon dust.

On the other hand, VDOT did not anticipate other fabricator-reported challenges that potentially raised costs, including the following:

- Maintenance of specified interpass temperatures during CJP welds.
- Level of effort involved to prepare the beveled surfaces for CJP welds.
- Quality shortcomings of web-to-flange fillet welds that required a change from the horizontal to the flat welding position.
- Girder mismarking that necessitated extra vigilance, as well as additional (stainless steel) fasteners.
- Environmental exposure of blasted girders after carbon dust exposure that necessitated 30–40 additional hours of blasting.
- A temporary absence of specified consumable welding material compliant with federal Buy America rules that caused delay and unanticipated search costs.

Lastly, discussions with other agencies with recent experience of ASTM A1010 in bridge girders or beams have suggested that significant savings may be achieved in future projects simply by standardizing the thickness of A1010 throughout a project, rather than using a variety of thicknesses as VDOT elected to do for this project.

Conclusions

Overall, the design, fabrication, and construction of the A1010 steel members used on the Route 340 bridge included many successes. The bridge was the first to use A1010 haunched girders, stainless steel bolts and fastener assemblies, and A1010 cross frames. The project showed that a conventional steel bridge fabricator can successfully fabricate a bridge from A1010 steel. Through detailed testing of stainless steel bolts, it was shown that all-stainless steel bolted splice connections can be achieved. All of the cross frames were formed from A1010 steel bent plates, which proved that the material can be successfully bent and fabricated by a bridge fabricator. Welding consumables were discovered to comply with the federal Buy America regulations.

The project also included a few lessons which could be applied to future A1010 bridge projects. VDOT officials discovered that the mill markings on the steel did not include the grade of steel, which is significant since the ASTM designation includes two grades. This observation was brought to the attention of the steel supplier. Subsequent A1010 bridges in Virginia will likely not be blast cleaned using garnet media because of the uneven formation of the patina between girders; it is anticipated that future girders will be left unblasted or will be blast cleaned using steel shot. The cost analysis revealed that cost savings for future A1010 bridge projects could be achieved through the use of uniform depth girders, rather than haunched girders because of the web waste associated with a haunched girder design and the increased cost of A1010 relative to typical carbon steels. Of course, this would not have been done on this project since the haunched girder design was necessary because of the site restrictions. Other areas in which cost savings can be achieved for future A1010 projects include using conventional steel cross frames and bolts rather than A1010 and stainless steel, respectively, conventional machine-blasting, and bridge design using fewer plate thicknesses than in this project.

Based on results of this project, it is recommended that additional research be conducted to evaluate the welding process and use of traditional NDT techniques for A1010 and the 309L filler metals. Another area of recommended additional research is investigation of A1010 as a repair material for existing bridges which are located in corrosive environments. This could include macro-environments such as nearness to the coastline or an industrial site, or micro-environments such as near beam ends under joints or localized areas, in particular close to water.

Acknowledgments

The authors are grateful for the cooperation of all VDOT personnel who contributed to the project. In particular, the authors thank Rex Pearce, Park Thompson, Eulogio Javier, Wayne Nolde, and Stephen Wright of the VDOT Staunton District. Bill Ordel and Selase Torkornoo of VTRC are recognized for their contributions to this project. Andrew Zickler of the VDOT Structure and Bridge Division and Thomas Darby of the VDOT Materials Division are also recognized for providing beneficial guidance during the project. The authors also thank members of the Federal Highway Administration, particularly Justin Ocel, Vanna Lewis, and Rodolfo Maruri.

Author Contributions

The authors confirm contribution to the paper as follows: study conception and design: Stephen R. Sharp, Jason Provines, Audrey K. Moruza, William F. Via, Jr. and Keith N. Harrop; data collection: Stephen R. Sharp, Jason Provines, Audrey K. Moruza, William F. Via, Jr. and Keith N. Harrop; analysis and interpretation of results: Stephen R. Sharp, Jason Provines, Audrey K. Moruza, William F. Via, Jr. and Keith N. Harrop; draft manuscript preparation: Stephen R. Sharp, Jason Provines, Audrey K. Moruza, William F. Via, Jr. and Keith N. Harrop. All authors reviewed the results and approved the final version of the manuscript.

References

1. Fletcher, F. B. *Improved Corrosion-Resistant Steel for Highway Bridge Construction*. FHWA Report No. FHWA-HRT-11-062, U.S. Department of Transportation, Washington, D.C., November 2011.
2. Seradj, H. *Oregon's Experience with ASTM A1010*. TRB Report No. 15-3926, Washington, D.C., 2015.
3. ArcelorMittal. *Plate: Duracorr®/A1010: Bridge Applications and Fabrication Guidelines*. https://iowadot.gov/bridge/A1010/presentations/arcelormittal%20duracorr%20bridges.pdf. Dated September 2013.
4. FHWA. *Technical Advisory: Uncoated Weathering Steel in Structures*. Technical Advisory 5140.22. https://www.fhwa.dot.gov/bridge/t514022.cfm. Accessed November 6, 2017.
5. Sharp, S.R., and A. K. Moruza. *Field Comparison of the Installation and Cost of Placement of Epoxy-Coated and MMFX 2 Steel Deck Reinforcement: Establishing a Baseline for Future Deck Monitoring*. VTRC 09-R9. Virginia Transportation Research Council, Charlottesville, 2009.
6. Williams, T. R., X. Xia, T. E. Darby, III, and S. R. Sharp. Evaluation of Stainless Steel Fasteners for Bolted Field Splice Connections of ASTM A1010 Corrosion-Resistant Steel Plate Girders. *Transportation Research Record: Journal of the Transportation Research Board*, 2017. 2630: 85–94.
7. American Society of Testing and Materials. A1010/A1010M. Standard Specification for Higher-Strength Martensitic Stainless Steel Plate, Sheet, and Strip. *Proc., Annual Book of ASTM Standards*, Vol. 01, No. 03, ASTM International, West Conshohocken, Pa., 2013.
8. Lincoln Electric Company. *Lincolnweld 309/309L Product Information Sheet*. http://www.lincolnelectric.com/assets/global/Products/Consumable_StainlessNickelandHighAlloy-Lincolnweld-Lincolnweld309309L/c61025.pdf. Accessed July 19, 2017.
9. Seradj, H. Oregon's First ASTM A1010 Bridge and Risk Management. *World 5 Steel Bridge Symposium*, Dallas, Tex., April 2012.

The Standing Committee on Fabrication and Inspection of Metal Structures (AFH70) peer-reviewed this paper (18-02966).

TRR
JOURNAL OF THE TRANSPORTATION RESEARCH BOARD

Article

Transportation Research Record
2018, Vol. 2672(26) 77–87

DOI: 10.1177/0361198118774162
journals.sagepub.com/home/trr
SAGE

A Decision Support Framework for Assessing the Contextual Factors for Complex Highway Projects

Akanksha Sinha[1], Kelly C. Strong[2], Mehmet E. Ozbek[3], and Jennifer Shane[4]

Abstract

Traditional strategies for highway projects during the 1950s and 1960s focused on three dimensions of project management: cost, schedule, and technical (scope). Recently, with the focus shifting towards reconstruction/rehabilitation projects, project management strategies have shifted to include other project management dimensions. The Second Strategic Highway Research Program (SHRP2) project R-10 examined the best practices for managing complex renewal projects. The primary outcome of the study was a recommendation to utilize a five-dimensional project management planning (5DPM) model that added context and financing as two new dimensions to the traditional dimensions of cost, schedule, and technical. Pilot testing of the 5DPM implementation suggested that the most complicated dimension to assess during the project management planning phase for a complex project is the context dimension. Currently there is no efficient, structured process for evaluating the context dimension on complex projects within the 5DPM framework. Given this, the objective of this study is to develop a decision support framework which can be used by different transportation agencies when assessing contextual factors and assigning complexity rating scores. The framework uses a structured analytical process as opposed to the more subjective scoring used during the 5DPM workshops. As a major part of its methodology, this research uses a multiple-criteria decision-making tool called analytical hierarchy process in developing the framework. The paper presents two implementation examples which demonstrate the feasibility of the developed framework for a rebuild project and an expansion project.

The number of transportation renewal projects has increased significantly, primarily due to aging infrastructure, increased traffic volume, and rising urbanization. Renewal projects for existing highway infrastructure play an essential role in society's economic, social, and political life and have a long-lasting effect on the community (*1*). In response to the deterioration of highway infrastructure, U.S. federal and state transportation agencies have shifted their focus from building new facilities to restoring, resurfacing, rehabilitating, and reconstructing existing ones (*2*, *3*). The construction industry does not have a universally accepted definition of complexity, nor does the transportation subset. Complexity is often uniquely defined based on an individual's or group's experiences and understanding of the task. Often there is a tendency to explain complexity in terms of a dollar value, however, there are many small projects that are complex, and this definition falls short (*4*). One current working definition is (from page 3 of (*4*)): "project complexity is the degree of differentiation of project elements, interrelatedness between project elements, and their consequential impact on project decisions." Regardless of the definition, transportation renewal projects are inherently complex owing to logistical challenges (e.g. phasing, maintenance of traffic, access), new construction methods, and restrictive regulations. Adding to the complexity is a demand to deliver the projects faster and with more control over time and cost (*5*–*8*). There has also been a decrease in the funds available to construct new roads or rebuild existing ones (*9*). The Federal Highway Trust Fund, which is the main source of funds for these projects, has been declining as its main source is the fuel tax, which has not changed since 1993 (*9*, *10*).

To address these problems, the Second Strategic Highway Research Program (SHRP2) R-10 project, "Project Management Strategies for Complex Projects", was carried out with the aim of developing effective

[1]Cornell University, Ithaca, NY
[2]University of Northern Iowa, Cedar Falls, IA
[3]Colorado State University, Fort Collins, CO
[4]Iowa State University, Ames, IA

Corresponding Author:
Address correspondence to Kelly C. Strong: kelly.strong@uni.edu

project management strategies and tools to address the challenges of managing complex transportation renewal projects (*5*). The results of the R-10 research introduced the five-dimensional project management (5 DPM) framework and identified various complexity factors within each of five dimensions (*5*). The managers of such complex projects have to optimize the available resources adhering to the technicalities of the projects and deal with the known and unknown constraints, and thus additional dimensions of financing (known) and context (unknown) were added to the traditional dimensions of cost, schedule, and scope (*5*). This multi-step approach, along with the development of a complexity mapping system, led to the development of five project development methods and 13 management tools that improve the project management of complex renewal projects (*5*).

The R-10 team also conducted several pilot workshops with different Departments of Transportation to test the efficacy of the 5DPM framework and refine the training material (*5*). Various participants were grouped together to form the project teams for the pilot workshops, and a two-step scoring system was used to map the complexity of the project. The first step consisted of ranking the five dimensions in order, from most complex to least complex (*5*). In the second step, the participants were asked to assign values from 0 to 100 to each of these dimensions based on their impact on the project (called the "the dimensional impact rating") (*5*). Although the complexity mapping developed by the research team provided an effective solution, it was not efficient in the sense that it took almost a day and half to complete the entire process (*5*).

During the pilot testing, it was noticed that the most difficult dimension to plan for while drafting the project management plan was the context dimension, which refers to the external factors (*5*). Given this, the purpose of this research is to develop a decision support framework which can be used by transportation agencies to assess the context dimension by using a structured, analytical approach. This research uses a multiple-criteria decision-making (MCDM) tool called analytical hierarchy process (AHP) in developing that framework.

Literature Review

A report published by the National Cooperative Highway Research Program identified a number of contextual factors that cause cost and schedule issues, including difficulties in obtaining rights-of-way, utility conflicts, underground conditions, and environmental and political issues. The report concluded that the impact of these issues can be decreased by using effective project management strategies (*5*, *11*).

Research on large infrastructure projects has identified the general public as one of the most important stakeholders and a major factor in the success of the project (*5*), primarily because of the economic and social challenges they face (*12*). These issues can also give rise to secondary effects, such as changing the demographics of the population, business loss, and growth inducement which can lead to social inequity (*5*, *13*). When multiple jurisdictions are involved in a highway project, legislative issues have the ability to influence the progression of a project and are one of the major causes of increase in project scope (*5*, *12*).

Martin and Does (*14*) concluded that a critical factor in the success of any project is its ability to minimize the impacts and inconvenience of traffic delays during the construction process (*14*). Public trust and confidence is important in an agency's ability to attract and invest infrastructure resources (*15*). Megaprojects or multifaceted projects face challenges such as greater transportation capacity, complex multimodal solutions, and designs sensitive to environmental context. They have impacts on a community's daily commerce, quality of life, and environment which makes it necessary for these projects to be as transparent as possible, and not to dismiss potential short-term negative impacts (*5*, *15*). Lee et al. (*16*) mentioned that the California Department of Transportation (Caltrans) used a constructability analysis tool to determine the most efficient construction strategies to minimize traffic delays and maximize the production capability (*16*). Many transportation renewal projects, especially intermodal projects, may require the relocation of existing utilities which represents a significant complexity issue affecting cost and schedule and requiring the cooperation of multiple groups (*5*, *17*, *18*).

The evidence from the literature clearly identifies the potential impact that contextual factors can have on project success. A structured, analytical process for identifying and quantifying the relative importance of contextual factors on a project would improve project management planning and resource allocation decisions. The decision support framework presented in the following sections provides a systematic method for transportation agencies to assign weights to contextual factors and rank them in the order of importance based on the priorities of their project. This framework also facilitates a time-efficient process for developing complexity factor rankings by the project team without the need for day-long workshops.

Methodology

The quantitative assessment of the factor categories and the factors that have been identified in the R-10 study is the primary objective of the proposed framework and, to achieve this, a four-step methodology was used by the research team. In the first step, the initial contextual factor categories and factors were identified based on the previous study by the SHRP2 R-10 team. After the identification of these factor categories and factors, meetings were conducted with the project management teams of the transportation agency to vet the factors identified from the literature based on their relevance to the project under investigation. The team members were asked to rate the factors on a scale of 0 to 2, where 0 indicated not relevant to the project, 1 indicated relevant, and 2 very relevant. Factors that received a mean score lower than 0.5 were removed from further consideration; and the factors that received a mean score of 1.5 and higher were retained in the list. The factors that received a mean score between 0.5 and 1.5 were discussed in detail by the members of the project management team until a consensus was reached on whether to retain or eliminate the factor. As suggested by Strong et al. (*19*), a scale of 0 to 2 was chosen as the goal was to vet the list quickly (by directly removing the factors with a mean score lower than 0.5 and retaining the ones with a mean score higher than 1.5) to free up valuable meeting time to focus on the remaining factors where disagreements may exist. At the end of this discussion, the team members were asked to identify additional factors that should be included in the subsequent list. The final list of contextual factors was then used to develop the MCDM decision support framework using AHP.

MCDM is a study of methods and procedures which provides a systematic and transparent approach to address the multiple and conflicting criteria in the decision process by enhancing objectivity and generating results which can be trusted with satisfaction (*20–22*). AHP was developed by Saaty in 1980 and is a multicriteria decision-making model which uses relative measurement on absolute scales of both tangible and intangible criteria based on the judgment of expert people (*20*, *23*). It uses multi-level hierarchical structure of alternatives in which the decision maker performs a simple pairwise comparison to generate the priorities (relative weights) of those alternatives (*24*).

In the first step of conducting AHP, the goal is broken down into a hierarchy of criteria, sub-criteria, and alternatives (*24*). This helps in establishing the relationship between the elements of all the levels in a networked manner (*24*). In the second step, a pairwise comparison is performed between the elements in the same hierarchy to determine the relative priority (weight) of each element in the hierarchy (*24*). The pairwise comparison mechanism uses a 1-to-9 scale in which a value of 1 means that both the elements have equal importance and contribute equally to the objective whereas a value of 3 means that the element that is selected has moderate importance and the experience and judgment of the decision maker slightly favors one element over the other (*24*). Similarly, a value of 5 indicates that one of element has strong importance; a value of 7 indicates very strong importance and a value of 9 indicates extreme importance (*24*).

To present the implementation of the decision support framework, two complex transportation projects were selected as case studies. The first project is the US-34 Permanent Rebuild project near Estes Park, Colorado, which was chosen to represent the requirements and priorities of a project for which the entire corridor needs to be closed during the construction of the project. US-34 is made complex by its unique funding (disaster relief special appropriation), its location in a narrow canyon with limited detours, and the need to maintain some level of access for emergency vehicles in the closed sections. The participants for this case study included four members of the project management team: two were representatives of the engineering consultant company for this project and the other two were representatives of the Colorado Department of Transportation (CDOT). The second project is the I-25 North Expansion project in Weld and Larimer Counties in Colorado, which was chosen to represent the complexity of a high priority, phased project built under traffic. The I-25 North Expansion project was made complex by the multi-jurisdictional funding pool, the temporary alignment designs (the current alignment is not exactly the same as the envisioned "permanent alignment" solution), and restricted limits of construction beyond the existing roadway. The participants for this case study included six members of the project management team: one member was from the Federal Highway Administration (FHWA) and the other five were from CDOT. For each project, one face-to-face meeting was conducted with the project management team and follow-up discussions were conducted via email. The following section presents the development of the AHP decision support framework for the two implementation cases.

Decision Support Framework Implementation

As mentioned, the first step of the MCDM implementation was to identify the contextual factor categories and factors which should be considered for a complex project. The R-10 team had identified eight factor categories and 26 factors under the context dimension. The definitions of the factor categories and factors under each category (which were identified through the SHRP2 R-10 study) are shown in Table 1 and were presented to the CDOT project management teams of the two case projects.

Table 1. List of Factor Categories with Definitions and Factors (from SHRP2 R-10)

Factor categories	Definition	Factors
Stakeholders	Everyone who is directly or indirectly associated with the project and potentially affected by the project in some way, including member of public, politician, owner, and jurisdiction.	- Public - Politician - Owner - Jurisdiction
Project-specific demands	This category includes all those factors that are directly related to the project, such as maintaining capacity, work-zone visualization, and intermodal requirements.	- Maintain capacity - Work-zone visualization - Intermodal requirements
Resource availability	This category includes all types of resources that may be required for the project and the capability of the project management team to gather all the resources in time.	- Availability of direct resources - Project management capabilities
Environmental	This category includes factors that affect the environment and the design limitations that are imposed on the project because of environmental regulations.	- Sustainability - Limitations
Legal and legislative requirements	This category relates to the legal and legislative requirements required when many parties are involved or when there is some legal restriction as per the existing laws to complete the project.	- Procedural laws - Local acceptance (local jurisdiction)
Global and national events	This category includes factors which affect the project owing to financial or political instabilities at the global or national level.	- Economics - Incidents
Unusual conditions	This category includes factors related to the conditions that are abnormal and unforeseen.	- Weather - Force majeure
Localized issues	This category includes factors that affect the public and businesses in the area where the construction is being carried out.	- Social equity - Integration of land use planning, growth inducement, and economic impact - Demographics - Land acquisition - Public emergency services - Work force issues - Utilities - Cultural - Marketing

After the identification of the initial list of factor categories and factors, each project management team customized each of these factor categories and factors based on its relevance to their project. Therefore, the next steps in the framework implementation are presented separately for each project.

US-34 Permanent Rebuild Project Implementation

The second step in the decision support framework was to conduct meetings with the members of the project management teams of both the projects to vet the existing factors, remove the non-relevant factors, retain the factors specific to the project, and add additional factors relevant to the project but not identified through the R-10 study. The participants were sent a document containing the definitions of the contextual factor categories and factors one week prior to the meeting so that they could familiarize themselves with the parameters of the factors to be discussed. Using the 0-to-2 vetting process described earlier, the US-34 team agreed on 14 factors as relevant while the remaining 12 factors were the subject of some disagreement and were refined in a facilitated discussion. The 12 factors discussed were: Intermodal Requirements, Direct Resources, Project Management Capabilities, Local Acceptance, Global and National Economics, Global and National Incidents, Force Majeure, Social Equity, Land Use, Growth Inducements and Economic Impacts, Demographics, Workforce Issues, and Utilities. After a detailed discussion, out of the 12 factors, nine were excluded and three were included. The participants were asked to add additional factors that may not have been included in the R-10 list, but they did not identify any additional factors. Thus, for the US-34 Rebuild team, a total of 17 factors and seven factor categories were finalized for the next step, as shown in Table 2.

The factors: politician, owner, jurisdiction, maintaining capacity, land acquisition, public emergency services, and marketing each received a perfect mean score of 2. Even though they received a perfect score, they may not be the most important factors and the score merely represents majority consensus among the participants for the inclusion of these factors.

Table 2. Final List of Factor Categories and Factors from US-34 Rebuild Implementation

Factor categories	Factors	Average score	Selected (Yes/No)
Stakeholders	Public	1.75	Yes
	Politician	2	Yes
	Owner	2	Yes
	Jurisdiction	2	Yes
Project-specific demands	Maintaining capacity	2	Yes
	Work-zone visualization	1.75	Yes
	Intermodal requirements	0.75	No
Resource availability	Direct resources	1.25	No
	Project management capabilities	1	No
Environmental	Sustainability	1.5	Yes
	Limitations and constraints	1.5	Yes
Legal and legislative	Procedural laws	1.5	Yes
	Local acceptance	1	No
Global and national events	Economic factors	1	Yes
	Incidents	0.75	Yes
Unusual conditions	Weather	1.5	Yes
	Force majeure	0.75	No
Localized issues	Social equity	1.25	No
	Land use planning, growth inducement, and economics	1	No
	Demographics	0.75	Yes
	Land acquisition	2	Yes
	Public emergency services	2	Yes
	Workforce issues	0.5	No
	Utilities	0.75	No
	Cultural factors	1.75	Yes
	Marketing	2	Yes

Table 3. Pairwise Comparison for Stakeholder by One Participant for US-34 Rebuild Project

Factor A	Factor B	More important factor	Degree of importance
Public	Politician	B	7
Public	Owner	A	2
Public	Jurisdiction	B	7
Politician	Owner	A	8
Politician	Jurisdiction	A	4
Owner	Jurisdiction	B	9

The third step of the framework was to develop and distribute the survey instrument based on the AHP MCDM methodology. The survey instrument was sent to the same participants of the US-34 project management team with whom the meeting was conducted to maintain consistency of results. In the survey, the respondents were asked to perform pairwise comparison first within each category and then between the categories. As an example: within the Stakeholder category, which had four factors, the participants were asked to compare each factor with another one to identify which one was more important and to assign magnitude on a scale of 1 to 9 as per the AHP methodology. The pairwise comparison results for Stakeholders by one of the participants is shown in Table 3. To illustrate, in the first row of Table 3, the participant compared Public (Factor A) and Politician (Factor B) and judged Factor B to be more important by a degree of 7 (where 7 means very strong importance).

The fourth and final step of the decision framework was to combine the results obtained from each individual member and perform the group AHP to develop one overall ranking as a group for all the factors and factor categories. The individual decisions were combined into group decisions for which the geometric mean of the individual decision was used to acquire the group decision for each pairwise comparison (*25*). After this, AHP analysis was performed by calculating the product, nth product, and priority vectors (weights) for each factor and factor category to determine the final weights. In Table 4,

Table 4. Global Vector (Final Weights) for Each Factor for US-34 Rebuild Project

Factors	Local priority vector for factors	Priority vector for factor categories	Global vector (final weight for each factor)
Stakeholders			
Public	0.21	0.22	0.05
Politician	0.22	0.22	0.05
Owner	0.29	0.22	0.06
Jurisdiction	0.28	0.22	0.06
Project-specific demands			
Maintaining capacity	0.36	0.14	0.05
Work-zone visualization	0.64	0.14	0.09
Environmental			
Sustainability	0.40	0.11	0.04
Limitations and constraints	0.60	0.11	0.07
Global and national events			
Economic factors	0.43	0.06	0.03
Incidents	0.57	0.06	0.03
Localized issues			
Demographics of population	0.21	0.30	0.06
Land acquisition	0.33	0.30	0.10
Public emergency services	0.14	0.30	0.04
Cultural factors	0.07	0.30	0.02
Marketing and public relations issues	0.25	0.30	0.08
Legal and legislative requirements (procedural laws)	1	0.10	0.10
Unusual conditions (extreme weather)	1	0.07	0.07
Total			1

the results of the priority vectors for the factors and the factor categories are shown in the second and third columns respectively. To illustrate, the priority vector of 0.21 for Public represents the influence of the Public factor on the overall category of the stakeholders which means that the weight or the importance of Public is 21% of the overall category of Stakeholders. The most important factor as a group in the Stakeholder category is Owner with a weight of 29% (0.29) and the least important factor is Public with 21%.

An important aspect of the AHP method is to check for consistency of the responses. For this purpose, AHP defines the consistency ratio (CR) and explains how it can be calculated. For group AHP, the CR for each category and across category comparisons was less than 10%, which is the CR suggested by the AHP literature (*24*). Similar steps were performed for the other four categories which have more than one factor: Project-specific Demands, Environmental, Global and National Events, and Localized Issues.

Pairwise comparisons were also performed between categories to establish rank order importance of the categories. The most important category as a group, based on pairwise comparison between categories, was Localized Issues with a weight of 30% and the least important category was Global and National Events with a weight of 6%.

The global vector for the factors can be calculated by multiplying the local priority vector for that factor by the local priority vector for the category to which the factor belongs. For example, the global vector for Public can be calculated by multiplying 0.21 (priority vector for Public factor) by 0.22 (priority vector for Stakeholder category), which is equal to 0.05. Global vectors for each factor are shown in Table 4.

I-25 North Expansion Project Implementation

Similar to the previous implementation effort, the second step was conducted and the same 0-to-2 scoring protocol was followed which led to elimination of one factor, the retention of 14 factors, and discussion on the remaining 11 factors. The 11 factors discussed were: Direct Resources; Sustainability; Local Acceptance; Global and National Economics; Global and National Incidents; Weather; Force majeure; Social Equity; Land Use, Growth Inducement and Economic Impacts; Demographics; and Workforce Issues. A facilitated discussion was carried out on each of these factors and, out of the 11 factors, six were excluded and five were included. At this point, the participants were asked to add any additional factors which were not a part of the initial list. They mentioned that phasing and constructability was an issue for the I-25 project, as it carries a high volume of traffic and no part of the corridor could be entirely closed for construction. Also, one of the most important issues was expanding the highway capacity while causing minimum delay to the traveling public, so

Table 5. Final List of Factor Categories and Factors from I-25 North Expansion Implementation

Factor categories	Factors	Average score	Selected (Yes/No)
Stakeholders	Public	2	Yes
	Politician	2	Yes
	Owner	2	Yes
	Jurisdiction	2	Yes
Project-specific demands	Maintaining capacity	2	Yes
	Work-zone visualization	2	Yes
	Intermodal requirements	1.67	Yes
	Phasing and constructability	-	Yes
Resource availability	Direct resources	1.33	Yes
	Project management capabilities	1.83	Yes
Environmental	Sustainability	1	Yes
	Limitations and constraints	2	Yes
Legal and legislative	Procedural laws	2	Yes
	Local acceptance	1.16	No
Global and national events	Economic factors	1	Yes
	Incidents	0.67	No
Unusual conditions	Weather	1	No
	Force majeure	0.67	No
Localized issues	Social equity	0.83	No
	Land use planning, growth inducement, and economics	1.16	Yes
	Demographics	0.5	No
	Land acquisition	1.67	Yes
	Public emergency services	1.67	Yes
	Workforce issues	1.33	Yes
	Utilities	2	Yes
	Cultural factors	0.33	No
	Marketing	2	Yes

this factor was added under the Project-specific demands factor category as the fourth factor. It was defined as the phasing of the entire project to cause minimum inconvenience or disruption to the moving public. Thus, after all the discussion, the I-25 team identified 20 factors and seven factor categories for the next step of the framework, as shown in Table 5.

Similar to the US-34 implementation, pairwise comparisons were conducted for each factor and the categories; and then group AHP analysis was performed to determine the weights.

The most important category as a group, based on pairwise comparison between categories, was Environmental with a weight of 21% and the least important category was Global and National Events with a weight of 5%. Global vector results for each factor are shown in Table 6.

Discussion

The weights of the factor categories can be analyzed with respect to the complexity that each of the projects faced owing to construction restrictions. For the US-34 project, localized issues were the most important category with a relative weight of 30%. This result is logical, because US-34 is not a primary freight or commuter route and the entire highway was closed for construction causing business disruption and local access challenges in the area. Thus information sharing and communication with local businesses, land owners, and residents living in the neighboring areas was a critical project complexity that needed to be addressed. As for I-25, the most important issue was the environmental category, with a relative weight of 21%, because the alignment of the highway had to be changed which caused design and construction complexities for the project team. The least important category for both the project teams was Global and National Events, with a 6% weight for US-34 Rebuild Project and 5% weight for I-25 North Expansion project. Table 7 shows the comparison between factor categories for the US-34 Rebuild Project and I-25 North Expansion project.

Table 8 shows the individual factors with the highest weight for US-34 and I-25 projects. When the global weights for the factors were compared, it was found that the most important factor for US-34 was procedural laws and land acquisition and both had a relative weight of 10%. The next important factor for US-34 was work-zone visualization with a relative weight of 9%. Because US-34 represented a permanent rebuild of highway sections destroyed by the Big Thompson Canyon flood in 2013, the administrative regulations concerning sources

Table 6. Global Vector (Final Weights) for Each Factor for I-25 North Expansion Project

Factors	Local priority vector for factors	Priority vector for factor categories	Global vector (final weight for each factor)
Stakeholders			
Public	0.31	0.16	0.05
Politician	0.14	0.16	0.02
Owner	0.36	0.16	0.06
Jurisdiction	0.19	0.16	0.03
Project-specific demands			
Maintaining capacity	0.34	0.16	0.05
Work-zone visualization	0.13	0.16	0.02
Intermodal requirements	0.09	0.16	0.01
Phasing and constructability	0.44	0.16	0.07
Resource availability			
Availability of direct resources	0.39	0.13	0.05
Project management capabilities	0.61	0.13	0.08
Environmental			
Sustainability	0.27	0.21	0.06
Limitations and constraints	0.73	0.21	0.15
Localized issues			
Integration of land use planning, growth inducement, and economic impacts	0.06	0.11	0.01
Land acquisition	0.26	0.11	0.03
Public emergency services	0.13	0.11	0.02
Workforce issues	0.13	0.11	0.02
Utilities and railroad issues	0.33	0.11	0.04
Marketing and public relations issues	0.09	0.11	0.01
Legal and legislative requirements (procedural laws)	1	0.18	0.18
Global and national events (global and national economic factors)	1	0.05	0.05
Total			1

Table 7. Category Weights for US-34 Rebuild Project and I-25 North Expansion Project

Factor categories	US-34 rebuild project	I-25 North expansion project
Stakeholder	0.22	0.16
Project-specific demands	0.14	0.16
Environmental	0.11	0.21
Resource availability	NA	0.13
Global and national events	0.06	0.05
Legal and legislative requirements	0.10	0.18
Unusual conditions	0.07	NA
Localized issues	0.30	0.11

of funds were complex. Some funds were restricted to restoration of previous assets, and could not be used for "improvements" in the highway. Furthermore, the flood destroyed a number of structures (e.g. culverts, dams, and bridges) that provided access to houses which were also destroyed and could not be replaced. Lastly, the Big Thompson River had a different flow line post-flood, so many of the land use and land ownership issues were complex. The fact that US-34 would be closed during construction made work-zone visualization and communication a complex issue considering that alternate routes in canyon terrain are problematic.

Similarly, for the I-25 team, Procedural Laws was the most important factor with a relative weight of 18%. However, the next important factor was Limitations and Constraints with a relative weight of 15% followed by Project Management Capabilities with a relative weight of 9%. The Procedural Laws factor was rated so highly for both the projects as the funding was provided by many different parties; and each of these parties wanted some control over the design decisions, which caused complexity for the project management teams. The complexities involving Limitations and Constraints on the I-25 project were related to the requirements to maintain access and mobility during construction while working within a very constrained corridor. Interstate-25 in Weld and Larimer Counties is a major freight and commuter corridor which also provides access to the rapidly growing population centers in Loveland, Fort Collins, Greeley, and the surrounding towns. Rapid development near the corridor has constrained alignment alternatives and has made construction staging a challenge.

Table 8. Factors with Highest Weights for US-34 Rebuild Project and I-25 North Expansion Project

US-34 Rebuild project		I-25 North expansion project	
Individual factors	Weights	Individual factors	Weights
Procedural laws	0.10	Procedural laws	0.18
Land acquisition	0.10	Limitations and constraints	0.15
Work-zone visualization	0.09	Project management capabilities	0.09

The most striking difference between the factor weights for both the projects was that the weights were more evenly distributed between factors for US-34, whereas for I-25 few factors had very high weights and others had very low weights. This is consistent with the complexity mapping framework of the original R-10 study. Projects can be complex in different ways, and understanding the nature of the complexity is key to effective project management planning and resource allocation decisions. While both projects have administrative complexity that warrants a large project team, US-34 could benefit from access to specialized knowledge in work-zone visualization and local access and public information programs. The I-25 North project would benefit from early involvement of team members with construction expertise and phased scheduling to better manage the complexities facing the project team.

The AHP approach to decision support allows the transportation agencies to weigh the categories and factors differently through a structured analysis of their unique needs as shown in the two implementation examples. The group AHP approach was robust, with CRs below the suggested 10% for all the categories and the factors. One limitation of the study was that some of the factor categories, such as Legal and Legislative Requirements for both US-34 and I-25 projects, Unusual Conditions for the US-34 project, and Global and National Events for the I-25 project, had only one factor. This eventually led to a higher global weight for that factor. However, as the results are in concordance with the views presented in meetings with both the project teams, it provides a strong indication that the methodology was accurate and can assist in the assessment of contextual factors as sources of complexity. Thus, the AHP tool generates a MCDM decision support framework which allows the project management team to weigh the factors reliably in a manner adaptable to the priorities of each project while providing a structured logic underlying the decision analysis.

Conclusions and Future Research

The goal of this research was to develop a framework for the project management teams of various transportation agencies to assess the contextual difficulties for a complex highway project in a structured, time-efficient manner. Along the lines of the dimensional impact rating developed by the SHRP2 R-10 team, a need was identified to rate the factors under those dimensions, instead of just rating the dimension, by using a structured approach. Given the scope and time constraints, the weighting of factors was conducted only for the context dimension as it was identified from the literature that assessing contextual difficulties at the early stages of the project is essential to keep the project on schedule and within budget. Also, it was established through the literature that gaining public trust and support (both of which relate to the context dimension as defined in the R-10 study) is one of the most essential factors in the success of any project.

As seen from the results, these relative weights can help the project team to focus on complexity factors with higher weights. Furthermore, the project team can further discuss and eliminate the factors which have lower weights. For federal-aid projects, the FHWA requires agencies to draft a risk management plan before the funds can be released. While not all complexity factors are a risk, one of the ways in which these weights can be used is to aid in developing a risk registry and a structured risk management plan as part of the project management plan required on complex transportation projects. The complexity factors should be reviewed in the structuring and risk identification steps of the risk management process (*26*). For the factors deemed to be risks, the factor weights can be used as proxies in establishing initial risk assessments which are then integrated into the risk analysis and risk management plan. Also, with the help of these weights, the project management team can be more aware of issues that might cause delays and can develop a more realistic budget and schedule, with a project management team whose skill set matches the needs of the project. It is important to note that an agency does not have to have an AHP software to be able to implement the framework described in this paper. The pairwise comparisons and the factor weight computations can be performed using a spreadsheet program. Owing to the group AHP approach used in this research, the CR was within the required limit, which validates the results.

The SHRP2 R-10 team had identified two new dimensions that should be considered while developing a project management plan: context and financing, along with the existing three dimensions of cost, schedule, and technical. The team identified other factors and factor categories in each of these other four dimensions. This framework can be modified and similar methodology can be performed for the other four dimensions as well. Comparisons of all the factors across all the dimensions could lead to valuable results.

In addition, while working on this project, it was identified that some of the factors have overlapping definitions. In a future study, these factors can be reinvestigated to develop fewer factors and factor categories; and the definitions could be broadened.

Author Contributions

The authors confirm contribution to the paper as follows: study conception and design: Akanksha Sinha, Kelly Strong, and Mehmet E. Ozbek; data collection: Akanksha Sinha, Kelly Strong, and Mehmet E. Ozbek; analysis and interpretation of results: Akanksha Sinha, Kelly Strong, and Mehmet E. Ozbek; draft manuscript preparation: Akanksha Sinha, Kelly Strong, Mehmet E. Ozbek, and Jennifer Shane. All authors reviewed the results and approved the final version of the manuscript.

References

1. Shang, J. S., Y. Tjader, and Y. Ding. A Unified Framework for Multicriteria Evaluation of Transportation Projects. *IEEE Transactions on Engineering Management*, Vol. 51, No. 3, 2004, pp. 300–313.
2. Lee, E.-B., and D. K. Thomas. State-of-Practice Technologies on Accelerated Urban Highway Rehabilitation: I-15 California Experience. *Journal of Construction Engineering and Management*, Vol. 133, No. 2, 2007, pp. 105–113.
3. Herbsman, Z. J., W. T. Chen, and W. C. Epstein. Time Is Money: Innovative Contracting Methods in Highway Construction. *Journal of Construction Engineering and Management*, Vol. 121, No. 3, 1995, pp. 273–281.
4. Dao, B., S. Kermanshachi, J. Shane, S. Anderson, and E. Hare. Exploring and Assessing Project Complexity. *ASCE Journal of Construction Engineering Management*, Vol. 143, No. 5, 2017, p. 04016126.
5. Shane, J., D. Gransberg, and K. Strong. *Project Management Strategies for Complex Projects*. SHRP-2 Report S2-R10-RW-1. Transportation Research Board, Washington, D.C., 2014.
6. Sillars, D. Development of Decision Model for Selection of Appropriate Timely Delivery Techniques for Highway Projects. *Transportation Research Record: Journal of the Transportation Research Board*, 2009. 2098: 18–28.
7. Gransberg, D., J. E. Koch, K. R. Molenaar, and M. C. Loulakis. *Preparing for Design-Build Projects: A Primer for Owners, Engineers, and Contractors*. American Society of Civil Engineers, Reston, VA, 2006.
8. Puerto, C. L., D. Gransberg, and J. Shane. Comparative Analysis of Owner Goals for Design/Build Projects. *Journal of Management in Engineering*, Vol. 24, No. 1, 2008, pp. 32–39.
9. Kirk, R. S., J. Frittelli, L. Luther, W. J. Mallett, and D. R. Peterman. *Surface Transportation Funding and Programs Under MAP-21: Moving Ahead for Progress in the 21st Century Act* (P.L. 112-141). Congressional Research Service, Washington, D.C., 2012.
10. ASCE. *2017 Infrastructure Report Card*. American Society of Civil Engineers, Reston, VA, 2017.
11. Jacobs Engineering Group, PMSJ Resources, and Virginia Tech. *NCHRP Web-Only Document 137: Guidance for Transportation Project Management*. TRB, National Research Council, Washington, D.C., 2009.
12. Hertogh, M., S. Baker, P. L. Staal-Ong, and E. Westerveld. *Managing Large Infrastructure Projects: Research on Best Practices and Lessons Learnt in Large Infrastructure Projects in Europe*. AT Osborne BV, Netherlands, 2008.
13. Barnes, G., and S. Erickson. *Developing a Simple System for Public Involvement Conflict Management*. Minnesota Department of Transportation, Saint Paul, MN, 2006.
14. Martin, C., and J. P. Does. Accelerating Highway Bridge Demolition: An Innovative Approach. Paper presented at the 2005 Annual Conference of the Transportation Association of Canada, Calgary, Canada, 2005.
15. Capka, J. R. Megaprojects–They are a Different Breed. *Public Roads*, Vol. 68, No. 1, 2004, pp. 2–9.
16. Lee, E., C. Ibbs, J. Harvey, and J. Roesler. Construction Productivity and Constraints for Concrete Pavement Rehabilitation in Urban Corridors. *Transportation Research Record: Journal of the Transportation Research Board*, 2000. 1712: 13–22.
17. Broadhurst, J. V. From Highways to Skyways and Seaways-The Intermodal Challenge. *Public Roads*, Vol. 68, No. 1, 2004, pp. 28–33.
18. Crichton, D., and K. Llewellyn-Thomas. *F.G. Gardiner Expressway Dismantling Project from the Don Roadway to Leslie Street*. Presented at 2003 Annual Conference of the Transportation of Canada, St. John's, Newfoundland and Labrador, 2003.
19. Strong, K. C., M. E. Ozbek, A. Sharma, and D. Akalp Decision Support Framework for Transit-Oriented Development Projects. *Transportation Research Record: Journal of the Transportation Research Board*, 2017. 2671: 51–58.
20. Zardari, N. H., K. Ahmed, S. M. Shirazi, and Z. B. Yusop. *Weighting Methods and their Effects on Multi-Criteria Decision Making Model Outcomes in Water Resources Management*. Springer, New York, 2015.
21. Janssen, R. On the Use of Multi-Criteria Analysis in Environmental Impact Assessment in the Netherlands. *Journal of Multi-Criteria Decision Analysis*, Vol. 10, No. 2, 2001, pp. 101–109.
22. Macharis, C., A. Verbeke, and K. De Brucker. The Strategic Evaluation of New Technologies through Multicriteria Analysis: The ADVISORS Case. *Research in Transportation Economics*, Vol. 8, 2004, pp. 443–462.

23. Saaty, T. L. Decision Making with the Analytic Hierarchy Process. *International Journal of Services Sciences*, Vol. 1, No. 1, 2008, pp. 83–98.
24. Saaty, T. L., and L. G. Vargas. *Models, Methods, Concepts Applications of the Analytic Hierarchy Process*. Springer Science & Business Media, New York, 2012.
25. Saaty, T. L. Group Decision Making and the AHP. In *The Analytic Hierarchy Process* (B. L. Golden, E. A. Wasil, and P. T. Harker, eds.), Springer, New York, 1989, pp. 59–67.
26. Golder Associates Inc., Molenaar, K. R., M. Loulakis, and T. Farragut. *Guide for the Process of Managing Risk on Rapid Renewal Projects*. Transportation Research Board, Washington, D.C., 2014.

The Standing Committee on Construction Management (AFH10) peer-reviewed this paper (18-00876).

TRR
JOURNAL OF THE TRANSPORTATION RESEARCH BOARD

Article

Transportation Research Record
2018, Vol. 2672(26) 88–95

DOI: 10.1177/0361198118798484
journals.sagepub.com/home/trr
SAGE

Timing of Cost Certainty in Highway Construction Project Delivery: Perceptions versus Empirical Results

Arthur L. C. Antoine[1] and Keith R. Molenaar[2]

Abstract

The early timing of cost certainty during highway project delivery is a primary reason for using alternative contracting methods, but the relative performance of these delivery methods has not been quantified by previous researchers. This research addresses the gap in knowledge by quantitatively showing how the traditional design-bid-build delivery method and the alternative contracting methods of construction manager/general contractor (CM/GC) and design-build (D-B) compare in the timing of cost certainty on U.S. highway construction projects. A novel comparison of the performance of CM/GC versus D-B projects procured by best value is performed. The results reveal a difference between the perceptions of agency decision makers and empirical project performance. By comparing empirical project data with findings aggregated from experts in a Delphi study and with a review of the limited relevant literature, this paper highlights contradictory findings and inconsistencies in the relative performance among the different sources of information. For the alternative contracting methods, the empirical project results show earlier timing for cost certainty than perceived by agency experts. Results reveal that agency personnel underestimate the potential time savings of CM/GC because they believe that D-B is faster to the point of cost certainty. However, empirical data show that CM/GC is substantially faster. The authors emphasize the need for further research to clarify the contradictions and inconsistencies in order to produce reliable, validated results that can provide useful recommendations for application in highway construction project delivery.

Departments of transportation (DOTs) consistently seek to be efficient with the use of public funds.

As a result of aging infrastructure, limited financial resources, and increased public scrutiny of transportation projects, highway construction agencies are turning to alternative contracting methods to reap the potential benefits, particularly schedule compression (*1–3*). Aggressive schedule compression is an influential factor when selecting alternative contracting methods and leads to the early knowledge of a project's cost, which is vital information for managing both project and program expenditures. The point in time when cost certainty is achieved during project delivery is the point in time at which an agency obtains a reliable estimate of the cost to complete a project (*1–4*). The time at which an agency obtains cost certainty during project delivery coincides with the point in time at which an agency defines the project scope and releases the design for construction. At this point in time, the agency has come to a contractual agreement for the cost of the project and anticipates minimal cost growth (*5*). The timing of the point of cost certainty is effectively a schedule or duration metric and not a cost metric. In the design-bid-build (D-B-B) delivery method, the point of cost certainty coincides with the time when design is fully completed and the agency awards construction to a general contractor. The point of cost certainty in the construction manager/general contractor (CM/GC) delivery method is known after the cost for the last construction work package is agreed on. While CM/GC projects may have multiple work packages, the true cost is not known until the agency awards the final work package. In design-build (D-B), the timing of cost certainty coincides with the award of the contract to the design-builder.

[1]Shrewsberry, Denver, CO
[2]Construction Engineering and Management, Department of Civil, Environmental and Architectural Engineering, University of Colorado Boulder, Boulder, CO

Corresponding Author:
Address correspondence to Arthur L. C. Antoine: AAntoine@shrewsusa.com

As a project performance metric, the timing of cost certainty is much less cited in literature than the straightforward metrics that directly indicate cost and schedule project delivery performance. Nonetheless, the early timing of cost certainty is valuable to highway construction, especially considering the current political climate which has heightened attention to expenditure on infrastructure. The timing at which an agency achieves cost certainty during project delivery is an important metric for both project and program management because this metric can facilitate the optimal use of often limited capital (*6*).

Researchers have well-established that the two alternative contracting methods of D-B and CM/GC outperform the traditional D-B-B delivery method, particularly with regard to schedule performance (*7–10*). However, during project delivery, agencies should be cognizant of the importance of cost certainty in addition to focusing on other performance goals such as cost reduction or schedule compression (*11*, *12*). Previous research studies show that early timing of cost certainty is a benefit of CM/GC (*3*, *4*) and D-B (*1–4*) over the traditional D-B-B delivery method. Still, there is a gap in knowledge of the relative timing of cost certainty between the alternative contracting methods. Notably, this paper provides a novel comparison of the timing of cost certainty for CM/GC versus D-B for U.S. highway projects that does not currently exist in the literature.

In light of the importance of the timing of cost certainty to highway agencies, the objectives of this research are to examine how D-B-B, D-B and CM/GC affect this performance metric and to highlight contradictions between the actual performance of highway project delivery methods and the perceptions of industry users of these delivery methods. This is accomplished by contrasting results from the analysis of empirical project data with findings based on reliable experiential knowledge of highway project delivery through a Delphi study.

Data Collection and Research Methods

The authors executed a two-phase approach to accomplish the research objectives. First, empirical data were gathered and analyzed. Then, because sufficient empirical data were not available at the level of granularity required to fully explore the timing of cost certainty in the second phase of research, experiential knowledge was obtained from qualified agency representatives. The experiential knowledge was collected through a rigorous Delphi process. This two-phase approach allows the authors to fully explore the empirical data and experiential knowledge to discover and validate findings from multiple perspectives. It should be noted that in this paper the D-B projects are distinctly D-B projects procured by best value (D-B/BV) (i.e., projects procured using selection factors in addition to cost).

Empirical Data Approach

To compare delivery method performance, it is imperative to have similar projects to obtain useful results (*2*, *7*). To achieve justifiably more comparable pools of projects, the authors developed an appropriate method to match similar projects between CM/GC and D-B/BV based on projects with similar characteristics. The process used numerous iterations of down-sampling from a larger database of empirical project information to ascertain the distinct cost range in which the alternative contracting methods are more frequently used by DOTs. Therefore, this study analyzed a small cost pool of projects with award costs from $10M to $50M. In this cost pool, the authors selectively chose projects that have similar characteristics between CM/GC and D-B/BV. Table 1 shows the project characteristics for D-B-B, CM/GC, and D-B/BV projects between $10M and $50M. The projects have similar complexity ratings, with only one D-B-B project being rated as non-complex based on the complexity definitions shown in Table 2. The facility type for all projects is predominantly road works. With regard to project type, the D-B-B and CM/GC projects match perfectly and are mostly rehabilitation projects. There is an even split of projects between new and rehabilitation for D-B/BV.

Experiential Data Approach

The experiential data were acquired through a Delphi study in which experts judge the performance of the project delivery methods for categorized descriptions of highway construction projects. The Delphi technique is appropriate to meet the objective of this paper because agency representatives can synthesize their experience across multiple sizes and project characteristics where empirical data do not exist. Key features of the Delphi technique that make it advantageous as a method of obtaining expert opinion are anonymity, iteration, controlled feedback, and the statistical aggregation of group response (*14*).

Following guidance from Hallowell and Gambatese, the authors determined the parameters for the Delphi study in advance of executing the method (*15*). The planned, sequenced format of the rounds and the type of feedback provided to participants in each round of questioning during the Delphi study was as follows:

- Round 1. Initial questionnaire distributed.
- Round 2. Issue Round 2 questionnaire that shows participants the aggregate response from Round 1

Table 1. Project Characteristics for D-B-B, CM/GC, and D-B/BV Projects between $10M and $50M

	Complexity rating			Facility type		Project type	
	Most	Moderate	Non-complex	Road	Bridge	New construction	Rehabilitation
D-B-B (*n* = 10)	6	3	1	9	1	1	9
CM/GC (*n* = 10)	6	4	0	9	1	1	9
D-B/BV (*n* = 10)	6	4	0	6	4	5	5

Table 2. Complexity Definitions (*13*)

Most complex (major) projects	Moderately complex projects	Non-complex (minor) projects
• New highways; major relocations • New interchanges • Capacity adding/major widening • Major reconstruction (4R; 3R with multiphase traffic control) • Congestion management studies are required • Environmental impact statement or complex environmental assessment required	• 3R and 4R projects that do not add capacity • Minor roadway relocations • Non-complex bridge replacements with minor roadway approach work • Categorical exclusion or non-complex environmental assessment required	• Maintenance betterment projects • Overlay projects, simple widening without right-of-way (or very minimum right-of-way take), little or no utility coordination • Non-complex enhancement projects without new bridges (e.g., bike trails) • Categorical exclusion

Note: 3R = resurfacing, restoration, rehabilitation; 4R = new construction/reconstruction.

and in which there is consensus for each judgment. Solicit feedback as remarks or reasons for outlying responses or changed responses. The participants should be shown their responses from the previous round.

- Round 3. Issue questionnaire but solicit input only for judgments with no consensus or stability after Round 2. Show the aggregate response from Round 2 along with the remarks obtained from Round 2. Solicit new remarks and show the participants their responses from the Round 2.
- Any necessary subsequent rounds. Issue questionnaire but solicit input only for judgments with no consensus or stability after the previous round. Show the aggregate response from the previous round along with the remarks. Solicit new remarks and show the participants their responses from the previous round.

In executing the Delphi method for this paper, respondents provided a response for individual judgments of performance for each alternative contracting method in relation to the performance of D-B-B, which was used as a baseline. Termination of judgments after each round of the Delphi was made on the basis of a measure of consensus and stability in the aggregate response (*16*). The question format for the Delphi method required a quantitative input for respondents' judgments of the performance of the CM/GC and D-B for a described project, relative to the provided D-B-B performance value. An example of a project description is, a "moderately complex, new, $15M" project. Respondents input an "X" if they would not recommend CM/GC or D-B for the described project. Therefore, respondents could provide a categorical variable and/or a continuous variable for each judgment in the Delphi. The aggregate response for the continuous variable is reported as the median value; for the categorical variable, the aggregate response is reported as the percentage of the respondents that input the categorical variable "X." Consensus and stability are assessed for each form of variable after each round during the Delphi study; the categorical and the continuous variables are respectively assessed by nonparametric and parametric methods.

Results and Analysis

In keeping with the two-phase approach to accomplish the research objectives, this section presents the results from the examination of the empirical data followed by the findings from the experiential approach. The limited number of projects in the empirical data set does not provide the level of granularity desired to fully explore the timing of cost certainty. Therefore, the experiential data were obtained for a further split of the cost range into

Table 3. Aggregate Information from the Empirical Data

Project delivery method	Award cost			Project duration		
	Mean (×$1,000)	Min. (×$1,000)	Max. award cost (×$1,000)	Mean (days)	Min. (days)	Max. (days)
D-B-B (*n* = 10)	$23,081	$11,429	$37,574	2,106	369	4,178
CM/GC (*n* = 10)	$23,912	$10,634	$39,600	662	204	1,059
D-B/BV (*n* = 10)	$18,604	$10,875	$43,960	1,420	643	3,647
Overall (*n* = 30)	$21,866	$10,634	$43,960	1,396	204	4,178

Note: min. = minimum; max. = maximum.

Table 4. Timing of Cost Certainty

Project delivery method	Timing of cost certainty			
	Mean (days)	Min. (days)	Max. (days)	Timing of cost certainty relative to D-B-B mean (days)
D-B-B (*n* = 10)	1,164	70	2,675	100%
CM/GC (*n* = 10)	329	55	649	28%
D-B/BV (*n* = 10)	765	71	2,858	66%
Overall (*n* = 30)	762	55	2,858	NA

Note: min. = minimum; max. = maximum; NA = not available.

project values between $10M and $20M and between $20M and $50M.

Empirical Results

In the empirical data, the average award cost for the projects from all project delivery methods is $21,866,000. The projects range in award costs from $10,634,000 to $43,960,000. Table 3 shows in aggregate the project cost and duration information from the empirical data. Contractor design costs are included for the D-B projects, whereas no design costs are included for the D-B-B and CM/GC projects. The National Highway Construction Cost Index by the FHWA is used to convert all costs to equivalent costs in June 2015. This conversion permits a fair comparison of project costs from different delivery methods.

Table 4 shows the results for the timing of cost certainty from the empirical data. The results show that, relative to D-B-B, project cost is known at least 66% earlier using either of the alternative contracting methods. This finding concurs with existing literature on the topic (*1–4*). Among the alternative contracting methods, CM/GC provides the earliest cost certainty. The superior performance of CM/GC in relation to D-B/BV provides new insight into the performance of these alternative contracting methods, and this finding is worthy of further exploration.

Experiential Approach: Delphi Respondent Characteristics

The approach taken in this paper is to supplement empirical data with experiential knowledge obtained from the Delphi method. The quality and consequential value of this experiential knowledge depends on the knowledgeability and expertise of participants (*14*, *17*, *18*). Participants in the Delphi study had to meet at least four out of five of the following selection criteria to participate in the Delphi study:

i. Has worked a minimum of 10 years in the highway design and construction industry.
ii. Works, or has worked, in an agency's alternative project delivery, contracts or procurements, or other relevant section that directly oversees or influences project delivery method selection.
iii. Has written papers or agency manuals on the topic of highway project delivery methods.
iv. Has a Professional Engineer (PE) licensure.
v. Has completed a highway project using at least two of the relevant project delivery methods.

The authors initiated the Delphi method by contacting a preliminary list of 30 agency personnel identified as potential expert participants. Ultimately, 18 experts were selected to participate in the Delphi study based on their qualifications and their interest and availability to

Table 5. Timing of Cost Certainty—Most Complex Projects

Timing of cost certainty relative to D-B-B				
Complexity	Most complex projects			
Project type	New projects		Rehabilitation projects	
Project size ($)	$10M–$20M	$20M–$50M	$10M–$20M	$20M–$50M
D-B-B baseline	100%		100%	
CM/GC	96%	100%	98%	100%
D-B/BV	90%	90%	90%	90%

participate. All of the 18 experts who started the Delphi study met four out of five requirements for the expert selection criteria. Fifteen of these experts met all five of the expert selection criteria. There was a 100% response rate for Round 1 from the 18 experts. The response rate for Round 2 was 94%, with only one of the 18 experts unable to participate. The response rate dropped to 78% for Round 3, with 14 respondents. These response rates are well within the norm for Delphi studies.

Of the 14 experts who completed the third and final round of the Delphi, 12 met all five of the expert selection criteria. In aggregate, the 14 experts who completed the Delphi study have an average of over 20 years' work experience in project delivery for highway design and construction projects; the minimum work experience is 9 years and the maximum is 34 years. All 14 of the experts work in an agency's relevant section or department that directly oversees or influences the selection of project delivery methods. Only two of the experts have not been involved in any publications on the topic of highway project delivery methods. Only one of the 14 participants does not have a PE license, but has an Engineer in Training certification. Collectively, the experts have worked on over 200 projects that used alternative contracting methods. All 14 experts have worked on D-B-B projects; two of the experts have not worked on CM/GC; and two experts have not worked on D-B/BV projects. These statistics attest to the exceptionally high quality of the expert panel for the Delphi study in this paper.

Another feature of the Delphi method for this paper that substantiates the quality of the expert participants is that all of the experts are from states that have experience in using the alternative contracting methods, CM/GC and/or D-B, based on the FHWA's EDC-2 Final Report (*19*). The experts' locations of employment are well distributed throughout the agencies in the U.S. that have significant experience in using alternative contracting methods; the locations/agencies include, Arizona, California, Colorado, Florida, Massachusetts, Michigan, Minnesota, Nevada, Utah, Wisconsin, FHWA Eastern Federal Lands Highway Division, FHWA Central Federal Lands Highway Division, and FHWA Western Federal Lands Highway Division.

Experiential Results

This section presents the results from the Delphi study along with explanations of the observed trends in the timing of cost certainty for the project delivery methods. The results represent the perceptions of the experts based on their collective experiential knowledge with specific regard to the timing of cost certainty. In other words, the results show trends of project delivery method performance in relation to the performance goal of achieving earlier cost certainty.

The results from the Delphi method reveal that the alternative contracting methods are recommended for projects of larger sizes and higher complexity. Therefore, only the results for most complex projects are shown in Table 5. The results highlight that experts believe that the alternative contracting methods achieve cost certainty earlier than D-B-B. However, it is apparent that in relation to D-B-B, CM/GC time savings are insignificant in the experts' eyes as they judge the timing of cost certainty to be similar to D-B-B. Results from the Delphi study show no consistent trend for D-B/BV relating to project type or size, but the results reveal that in aggregate the agency personnel believe D-B/BV has earlier timing of cost certainty than CM/GC.

Discussion

This section presents findings through a triangulation of empirical data, Delphi results (quantitative results and qualitative feedback), and relevant literature. The purpose of this paper is not to revisit the established superior performance of alternative contracting methods but to reveal new information with regard to achieving earlier cost certainty. This paper also seeks to highlight instances of contradictions among the multiple sources of information.

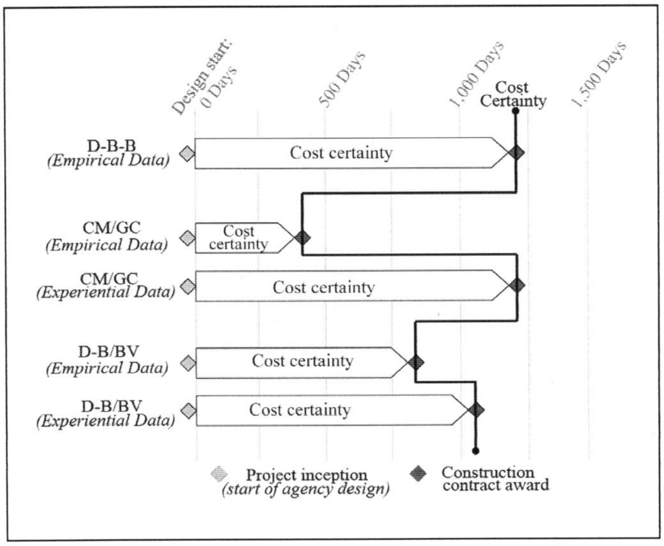

Figure 1. Summary of cost certainty findings—empirical versus experiential.

The results from the Delphi method suggesting that CM/GC and D-B/BV is not used on projects of lower costs and less complexity is supported by literature on the topic (*20–24*) and reinforced by the mutual perspective of the expert participants evidenced by comments received during the Delphi study, such as:

- "On projects of this size ($ 0–$10M) there is very little opportunity to obtain early cost certainty using D-B-B."
- "I believe alternative delivery will provide savings in time and money on larger projects."

Figure 1 presents a summary comparison of the timing of cost certainty that is to scale for the empirical versus experiential results. Figure 1 simplifies the alternative delivery processes for the purposes of this discussion. Specifically for CM/GC, the release of multiple work packages is not illustrated. The main contradiction between the empirical and experiential results observable in Figure 1 is:

- Empirical data: CM/GC timing of cost certainty < D-B/BV timing of cost certainty;
- Experiential data: CM/GC timing of cost certainty > D-B/BV timing of cost certainty.

The difference in perceived CM/GC versus empirical findings is stark. The expert perception of insignificant CM/GC time savings in relation to D-B-B warrants further investigation. This study cannot determine why the perceptions of agency personnel contradict the results established by literature and shown by the empirical data in this research. Previous qualitative literature asserts that the timing of cost certainty for CM/GC is earlier than D-B-B through obvious time savings achieved in the CM/GC delivery process. For instance, CM procurement occurs concurrently with design, therefore, there is essentially no letting period (*3*). There is less design effort required in CM/GC than in D-B-B and this can also offer time savings. With CM/GC, the designer need only produce a design that is constructible, not biddable; that is, the designer produces a design for one contractor and not a design that will be bid on by multiple contractors (*3, 4*).

There is also a pronounced difference in the perceived disparity between CM/GC and D-B cost certainty. There is limited literature available on the comparison of any schedule performance metric for CM/GC versus D-B. Existing publications that include both D-B and CM/GC rely on the experiential knowledge of industry users of these delivery methods to determine superior performance. Although none of these publications focus on the timing of cost certainty, most suggest that D-B is the superior time saving method (*9, 10, 25*). The results from the Delphi study concur with literature in this regard. However, there remains another discrepancy with the finding from this paper's empirical data that shows CM/GC to have a slightly earlier timing of cost certainty than D-B/BV. This discrepancy calls for further examination, especially as the quantity of the empirical data restricts the application of precise analytical methods to determine statistically significant differences.

The comments from the Delphi method elucidate agency personnel's thinking in relation to CM/GC versus D-B performance. It is apparent from the experts' qualitative feedback that agency professionals have opposing perspectives of the timing of cost certainty for CM/GC and D-B. Some agency professionals deem D-B to be the fastest and most practical method to get to the point of cost certainty based on the notion that D-B is the fastest among the delivery methods to facilitate a project as ready for bidding. Additionally, the qualitative feedback from the Delphi study hints that agency personnel believe the alternative contracting methods provide more reliable cost certainty in relation to final project costs. However, perspectives are mixed as to whether CM/GC or D-B provides the most reliable cost certainty. The following comments from the Delphi study highlight agency personnel's diverse ways of thinking in relation to CM/GC versus D-B timing of cost certainty:

- "I feel that D-B/BV would not reduce timing more than CM/GC due to a longer selection process."
- "Because you get to the full project contract stage earlier with the D-B method, you may finalize your costs earlier as well."

- "If this was to look at 'close enough/effective' cost certainty versus 'true/final' cost certainty, CM/GC would come out well ahead of D-B. We use CM/GC specifically for the purpose of early cost certainty."

The topic of reliable cost certainty wanders into the realm of other project performance metrics that are outside the scope of this research. This research did not include an examination of the reliability of cost certainty in the different project delivery methods, but this is certainly a topic for future research.

Conclusions and Scope for Future Research

The timing of cost certainty during highway project delivery has not been quantified by previous researchers. The results of this research paper address this gap in knowledge by quantitatively showing how the timing of cost certainty is affected by using different highway project delivery methods. Additionally, this paper provides a first-of-a-kind comparison of the timing of cost certainty for CM/GC versus D-B/BV projects, and highlights contradictions between the actual performance of highway project delivery methods and the perceptions of industry users of these delivery methods.

The authors explore empirical project data, supplement findings with information from the aggregate experiential knowledge of highway project delivery experts, and review the limited relevant literature on the timing of cost certainty. As a result, the authors highlight contradictory findings among different sources of information. For the alternative contracting methods, the empirical project results show earlier timing for cost certainty than perceived by agency experts. Among the alternative contracting methods, CM/GC provides the earliest cost certainty. Results reveal that agency personnel completely underestimate the potential time savings of CM/GC because they believe that D-B is the faster delivery method to get to the point of cost certainty when empirical data show the opposite.

It should be noted that both the empirical data and experiential data are susceptible to error. The empirical data pool is necessarily small in size because of the small number of completed CM/GC U.S. highway projects at the time of this study. The pool of experts for this study is undoubtedly qualified, but they also have relatively little experience with CM/GC because of the novelty of this delivery method. Finally, the literature is primarily based on qualitative results that could be subject to biases or a lack of complete information. These factors point to the need to conduct additional research to clarify any contradictions in the findings from multiple sources of information. The authors believe that the three-pronged Methodology presented in this paper is correct and produces reliable results that can be useful as recommendations for application in highway construction project delivery. However, the lack of CM/GC project experience across the U.S. presents an irreconcilable barrier. The study should be repeated when enough CM/GC projects are available to develop statistically reliable results. Also, in future work it would be beneficial to study award cost growth versus cost certainty timing and award cost growth versus project duration.

Acknowledgments

The authors extend their sincere gratitude to the numerous agency professionals who contributed to the collection of the empirical project data. The authors would also like to thank the DOT project delivery experts who obliged to participate in the Delphi study, which would have been unavailing without their expertise and time.

Author Contributions

The authors confirm contribution to the paper as follows: study conception and design: Arthur L. C. Antoine and Keith R. Molenaar; data collection: Arthur L. C. Antoine and Keith R. Molenaar; analysis and interpretation of results: Arthur L. C. Antoine and Keith R. Molenaar; draft manuscript preparation: Arthur L. C. Antoine and Keith R. Molenaar. All authors reviewed the results and approved the final version of the manuscript.

References

1. Songer, A. D., and K. R. Molenaar. Selecting Design-build: Public and Private Sector Owner Attitudes. *Journal of Management in Engineering*, Vol. 12, No. 6, 1996, pp. 47–53.
2. Shrestha, P. P., G. C. Migliaccio, G. E. Gibson, and J. T. O'Connor. Benchmarking of Large Design-Build Highway Projects: One-to-One Comparison and Comparison with Design Bid-Build Projects. *Transportation Research Record: Journal of the Transportation Research Board*, 2007. 1994: 17–25.
3. Gransberg, D. D., and J. S. Shane. *NCHRP Synthesis 402: Construction Manager-at-risk Project Delivery for Highway Programs*. Transportation Research Board of the National Academies, Washington, D.C., 2010.
4. Touran, A., D. D. Gransberg, K. R. Molenaar, and K. Ghavamifar. Selection of Project Delivery Method in Transit: Drivers and Objectives. *Journal of Management in Engineering*, Vol. 27, No. 1, 2011, pp. 21–27.
5. Xiao, H., and D. G. Proverbs. Cost Certainty and Time Certainty: An International Investigation. *Proc., Association of Researches in Construction Management ARCOM 19th Annual Conference*, Brighton, England, United Kingdom, 2003, pp. 23–31.

6. Hastak, D. M. Skills & Knowledge of Cost Engineering. *Proc., AACE International*, Morgantown, WV, 2015.
7. Konchar, M., and V. Sanvido. Comparison of US Project Delivery Systems. *Journal of Construction Engineering and Management*, Vol. 124, No. 6, 1998, pp. 435–444.
8. Gransberg, D. D., G. M. Badillo-Kwiatkowski, and K. R. Molenaar. Project Delivery Comparison Using Performance Metrics. *Proc., AACE International Transactions*, 2003, p. CS21.
9. McGraw Hill Construction. Project Delivery Systems: How They Impact Efficiency and Profitability in the Buildings Sector. *SmartMarket Report*, 2014.
10. Bingham, E., M. Asmar, and G. Gibson, Jr. Project Delivery Method Selection: Analysis of User Perceptions on Transportation Projects. *Construction Research Congress*, 2016, pp. 2110–2118.
11. Soetanto, R., D. G. Proverbs, and G. D. Holt. Achieving Quality Construction Projects Based on Harmonious Working Relationships: Clients' and Architects' Perceptions of Contractor Performance. *International Journal of Quality & Reliability Management*, Vol. 18, No. (4/5), 2001, pp. 528–548.
12. Del Puerto, C., E. Craigie, and D. Gransberg. Construction Cost Certainty versus Construction Savings: Which is the Correct Approach? Presented at 95th Annual Meeting of the Transportation Research Board, Washington, D.C., 2016.
13. Anderson, S., K. Molenaar, and C. Schexnayder. *NCHRP Synthesis 574: Guidance for Cost Estimation and Management for Highway Projects During Planning, Programming, and Preconstruction.* Transportation Research Board of the National Academies, Washington, D.C., 2007.
14. Rowe, G., and G. Wright. The Delphi Technique as a Forecasting Tool: Issue and Analysis. *International Journal of Forecasting*, Vol. 15, 1999, pp. 353–375.
15. Hallowell, M., and J. Gambatese. Qualitative Research: Application of the Delphi Method to CEM Research. *Journal of Construction Engineering and Management*, Vol. 136, No. 1, 2010, pp. 99–107.
16. Kalaian, S., and R. Kasim. Terminating Sequential Delphi Survey Data Collection. *Practical Assessment, Research & Evaluation*, Vol. 17, No. 5, 2012.
17. Skulmoski, G. J., F. T. Hartman, and J. Krahn. The Delphi Method for Graduate Research. *Journal of Information Technology Education*, Vol. 6, 2007, pp. 1–21.
18. Giannarou, L., and E. Zervas. Using Delphi Technique to Build Consensus in Practice. *International Journal of Business Science and Applied Management*, Vol. 9, No. 2, 2014.
19. Federal Highway Administration. *Every Day Counts.* EDC-2 Final Report. 2015, pp. 22–25, 69–72. https://www.fhwa.dot.gov/everydaycounts/reports/edc-2-finalreport/. Accessed June 10, 2015.
20. Gordon, C. M. Choosing Appropriate Construction Contracting Method. *Journal of Construction Engineering and Management*, Vol. 120, No. 1, 1994, pp. 196–210.
21. AECOM Consulting and Science Application International Corporation. *Design-Build Effectiveness Study – As Required by TEA-21 Section 1307(f).* US DOT Advisory Team slides, AECOM, Birmingham, AL, 2003.
22. Minchin, E., L. Ptschelinzew, G. C. Migliaccio, U. Gatti, K. Atkins, T. Warne, G. Hostetler, and S. Asiamah. *NCHRP Report 787: Guide for Design Management on Design-Build and Construction Manager/general Contractor Project.* Transportation Research Board of the National Academies, Washington, D.C., 2014.
23. Molenaar, K. R., and D. D. Gransberg. Design-builder Selection for Small Highway Projects. *Journal of Management in Engineering*, Vol. 17, No. 4, 2001, pp. 214–223.
24. Alleman, D., A. Antoine, M. Schrilla, and K. Molenaar. The Use and Performance of Alternative Contracting Methods on Small Highway Construction Projects. *The Ninth International Structural Engineering and Construction Conference (ISEC-9), Resilient Structures and Sustainable Construction*, Valencia, Spain, CPM-15, 2017.
25. Farnsworth, C., R. Warr, J. Weidman, and D. Mark Hutchings. Effects of CM/GC Project Delivery on Managing Process Risk in Transportation Construction. *Journal of Construction Engineering and Management*, Vol. 142, No. 3, 2015.

The Standing Committee on Construction Management (AFH10) peer-reviewed this paper (18-05987).

Gap in pagination due to unavailable paper.

Pages 96-105

Article

TRR
JOURNAL OF THE TRANSPORTATION RESEARCH BOARD

Transportation Research Record
2018, Vol. 2672(26) 106–113

DOI: 10.1177/0361198118797207
journals.sagepub.com/home/trr
SAGE

Effective Post-Construction Evaluation Practices for Work Zone Management Strategies

Michael J. Volk[1], Ryan S. Jeong[2], and Jennifer S. Shane[1]

Abstract

State departments of transportation (DOTs) invest a significant amount of effort to develop and implement transportation management plans (TMPs) in order to effectively manage the traffic flow of work zones and secure the safety of workers and the traveling public during construction of highway projects. State DOTs typically spend an average 3% to 5% of the total construction cost for TMP development and implementation, but some urban projects, in highly congested areas, may require up to 30% of the total construction cost for this effort. For continuous improvement of TMP development and implementation, it is critically important to monitor the performance of work zone management strategies during construction and evaluate them at the conclusion of construction. However, most state DOTs lack structured policies and/or work processes to ensure the implementation of these important monitoring and evaluation components as part of the TMP. A working feedback loop must be established in the project delivery process to document lessons learned, successes and failures, potential changes to the TMP, and transfer them back to the early TMP development stage. This research extensively reviews existing literature and evaluates the current practices of monitoring and evaluating work zone management strategies. This study identifies and presents effective practices for continuous improvement of TMP development and implementation through creating a working feedback loop.

The country's infrastructure is aging, and the demands placed on the roadways are continually increasing. As a result, state departments of transportation (DOTs) have been working on more rehabilitation and road replacement projects. This means that work zones are appearing more frequently on the nation's roadways. Proper planning and management is imperative for the success of these work zones; it is critical that they do not negatively affect the traveling public's safety and mobility or present significant health risks to the construction workforce.

The *Final Rule on Work Zone Safety and Mobility* was published on September 9, 2004, and mandated compliance by October 12, 2007 (*1*). This rule regulated that state and local governments receiving federal funds for roadway projects must begin to utilize transportation management plans (TMPs) to mitigate the effects of work zones in their vicinity and throughout the larger corridor of travel. Thus, state DOTs developed—and are continuing to develop—formal processes for designing, implementing, and evaluating TMPs. These formal guidelines and policies are within varying degrees of completion amongst the different state DOTs. Notably, the least developed process in most of the state DOTs investigated was the evaluation process. Evaluation can occur during TMP implementation, construction, or after completion of the construction phase. Evaluation has multiple scopes: at the narrower project-level, changes necessary during construction, or lessons learned following its completion are observed. At the broader agency level, a conglomeration of projects are assessed to determine necessary changes to the DOT's TMP policies and procedures.

Many state DOTs possess only a primitive evaluation process, if one exists. However, evaluation is extremely important because a formal feedback loop allows for continuous improvement of the TMPs, furthering the pursuit of work zone safety and mobility. This research scans current practices among the state DOTs, by reviewing DOT TMP manuals, and identifies best practices as a benchmark that may be a model for all state DOTs.

[1]Department of Civil, Construction, and Environmental Engineering, Iowa State University, Ames, IA
[2]University of Pennsylvania, Philadelphia, PA

Corresponding Author:
Address correspondence to Jennifer S. Shane: jsshane@iastate.edu

Overview of a TMP and Its Components

A TMP is a set of plans that are used to minimize the impacts the work zone has on the traveling public and to guarantee the safety of the construction workers. A TMP has three components: a Temporary Traffic Control Plan (TTC), a Public Information and Outreach Plan (PIOP), and a Transportation Operations Plan (TOP). The TTC is utilized to control the traffic that is inside and near the work zone: examples of its applicability include the use of temporary pavement markings and signs. The PIOP informs the public of important information like the position of the work zone, any detours, lane closures or expected delays—this may include radio broadcasts or newspaper advertisements. The TOP is for controlling the impacts of the work zone outside of the work zone and its immediate locale: examples include turn restrictions, ramp closures, and reversible lanes (*1*).

The Federal Highway Administration (FHWA) requires a TMP for all significant projects (*1*). A project is significant if, "alone or in combination with other concurrent projects nearby, [it] is anticipated to cause sustained work zone impacts that are greater than what is considered tolerable based on state policy and/or engineering judgement," or if it is taking place on the interstate system within a transportation management area (TMA) for three or more days with continuous or intermittent lane closures (*1*). A transportation management area is defined as, "an urbanized area with a population of over 200,000" or "any urbanized area for which TMA designation has been requested by both a state governor and the area's metropolitan planning organization and has been granted by the Secretary of the United States Department of Transportation" (*2*).

Most state DOTs have three levels of TMPs: major, moderate, and minor TMPs. A major TMP is for large-scale projects that are defined as "significant", such as replacing a bridge on an interstate inside of a TMA. These TMPs include a TTC, PIOP, and TOP. A moderate TMP is for a project that is not "significant" but is still expected to have notable work zone impacts that are beyond what is acceptable by the state's policies. An example of this would be repaving a heavily used road that is not an interstate. A moderate TMP has a TTC, and any applicable parts of the PIOP and TOP as needed. A minor TMP is for routine projects that will not cause a delay to the traveling public. An example is regular maintenance on a roadway, such as fixing potholes. A minor TMP consists of a TTC, with the PIOP and TOP being optional.

Literature Review

A literature review was conducted to determine the current practices of the design, implementation, and evaluation phases of TMP use for state DOTs. This consisted of reviewing thirty-four state DOT manuals for TMPs and work zone safety and mobility guidelines, the material made available by the FHWA online, and research performed by the Institute for Transportation (InTrans) at Iowa State University. The information in this paper comes from state transportation agency manuals.

Current Practices

The *Final Rule on Work Zone Safety and Mobility* was written in a manner that allows for flexibility so that the agencies are able to tailor the TMPs to their specific needs. "While it provides direction on areas that agencies need to address in their work zone policy, it does not provide specifics on what must be included and does not specify a particular format" (*1*). The result of the flexibility in the rule is that the state DOTs have developed their own unique standards and definitions for certain projects. This also allows the DOTs to define their TMPs in different ways. However, there are certain aspects of TMPs that are nearly identical across all DOTs and agencies. The rule of significance is practically universal—most agencies define it as a project that will cause significant impacts to the traveling public according to engineering judgement, or if the project involves continuous or intermittent lane closures for three or more days on an interstate that is within a TMA. Some projects might fall under the definition of "significant" because of where they happen to be geographically, but they might not have major work zone impacts on the traveling public; because of this it is possible for state DOTs to request an exemption from the FHWA.

Designation of Transportation Management Plans

Currently, most DOTs have three levels of TMPs, but due to the flexibility forwarded to the DOTs there is no formal nomenclature. New Mexico and South Carolina both use major, intermediate, and basic to define their TMPs (*3*, *4*). Major TMPs are for significant projects, whereas intermediate and basic TMPs are for nonsignificant projects (*3*, *4*). The California Department of Transportation (Caltrans) defines its TMPs as major, minor, and blanket. Caltrans uses its major TMPs for significant projects, whereas the minor and blanket TMPs are used for nonsignificant projects (*5*). Virginia and Delaware have similar designations for their TMPs, Delaware has type A and B, with type B for significant projects (*6*). Virginia TMP designations are type A, B, and C; type A is significant projects, while type B and C are for nonsignificant projects (*2*). Tennessee is straight forward with its naming of TMPs, significant TMPs are for significant projects, while nonsignificant TMPs are for nonsignificant projects, and exempt/recurring TMPs are for projects that will not have work zone impacts (*7*). Oregon DOT is different from the rest of the stated DOTs. Oregon has three TMPs: project, corridor, and program (*8*). Project-level TMPs are for single projects and is similar to the Type A and major TMPs used by Virginia and Caltrans (*8*). Corridor-level TMPs address how to handle the impacts of all the work zones along the corridor (*8*). The Program-level TMP is the framework for the corridor and program TMPs, and addresses traffic concerns across the state from all the work zones (*8*).

Creation of the TMP

When the DOT begins planning a project, they also begin to consider developing a TMP. As the project is developed, they assign a level of significance to it. After they have assigned the level of significance, they determine the TMP that will be needed to make the project safe and allow the road users mobility. After this, they hand the project to the development team which differs by DOT. Nevada has the project manager appoint a TMP team who is then responsible for its development (*9*). Colorado is on the other end of the spectrum; if the project is significant the contractor will develop the TMP (*10*). Based on interviews, Iowa is in the middle of the spectrum; for its major projects the contractor develops the TMP, all TMPs for non-major projects are developed by the districts.

Design and Implementation

Work on the TMP begins early in the planning phase for road work in order to determine the needs of the project and to minimize the cost of the TMP. The TMP is a living document, it is constantly updated and revised. This makes the initial planning of the document important, as the fewer changes that must occur later, the more money that is saved by the DOT. After the TMP and work zone have been designed, the next phase is the implementation. During this phase, there is typically a person placed in charge of monitoring if the construction is following the TMP. In Alaska, for example, the project engineer is responsible for ensuring that the TMP is being followed during construction, while a work zone traffic supervisor, from the contractor, is responsible for the implementation of the work zone and TMP (*11*). After this, the project engineer or another departmental employee must perform "daily night and day inspections of traffic control measures and correction of deficiencies" (*11*). Some DOTs do not have the resources to have a departmental employee inspect the work zone for compliance with the TMP. Iowa DOT relies on daily "safety diaries" filled out by the contractor to ensure TMP compliance.

Performance Measures, Monitoring and Revision of the TMP

After the TMP has been implemented, monitoring takes place in case there is a need for revisions to the TMP in order to facilitate faster travel or to make the work zone safer for workers or motorists. The state DOTs make use of performance measures to evaluate the work zone and effectiveness of the TMP. The performance measures are broken down into two categories: safety performance measures and mobility performance measures, see Table 1.

With these performance measures, the DOTs are able to determine if the TMP needs to be revised. The process for determining if a change is necessary varies across DOTs. Georgia monitors certain performance measures such as queue length, travel time, and volume, and then compares the actual recorded measurements against the predicted measurements to determine if the TMP needs to be revised (*13*). Illinois does not compare the actual measurements against the predicted, but they do perform queue analysis to improve motorist safety: this is done by predicting the largest queue that is possible so that the DOT can place "warning signs in advance of possible queues" (*14*). Colorado monitors the work zone weekly during peak and nonpeak times; measurements of work zone delay, and queue data are collected (*10*). How the TMPs are revised during construction differs across DOTs—some have formalized standards and procedures while other DOTs operate in a more informal manner. South Carolina has a more formal approach for revising their TMPs, using a transportation management plan team that meets to review reports from the resident engineers and to suggest changes to the TMPs so that they meet the required standards (*3*). A less formal approach is taken by Colorado, as they monitor their TMPs and if it is failing to meet the benchmarks that are required of it, they change it (*10*). The least formal approach that was seen during the scan was that of Iowa, where if the engineer in charge of the project decides that a change

Table 1. Performance Measures (*12*)

Safety performance measures	Mobility performance measures
Number of crashes	Queue length
Crash frequency/rates	Duration of queue
Worker casualties	Level of service
Inspection scores	Travel time
Work zone intrusion frequency	User complaints

needs to occur, they ask the district team that developed the TMP to change it.

Post-Project Evaluations

"Post project construction evaluation and performance assessments ... provide feedback into the project development and implementation loop. ... to identify promising work zone strategies, as well as areas for potential improvement" (*15*). Many DOTs differ regarding their post-project construction evaluations for three reasons. The first reason is that DOTs have different resources and needs. Some DOTs have not been able to dedicate the necessary resources to establish a formal feedback loop for post-project purposes. Second, it is not required to evaluate every TMP; some DOTs focus on significant projects or only evaluate projects that did not meet the standards. Finally, for DOTs that require a post-project evaluation, what is contained in the evaluations is different. Both Washington, D.C. and Wisconsin include in their post-project evaluations a portion that states how "useful" the TMP was or how useful the TMP evaluation method was (*16*, *17*). Virginia, Wisconsin, and Washington, D.C. all call for a summary of changes that were made to the TMP, the success of each change, and the public reaction to the TMP (*2*, *16*, *17*). These three DOTs, however, differ in their approach in terms of how they handle performance measures. Virginia requires a statement describing a summary of travel time and queue length during peak hours (*2*). Wisconsin requires a section in the TMP on the performance measures that were used during the project, not a standardized set of measures like Virginia requires (*16*). Washington, D.C. requires a discussion of the performance measures used. They compare these to the measurements that were taken against the predicted measurements (*17*). New Mexico DOT's post-project evaluation consists of the successes and failures of the TMP and any changes and their results (*4*). Most significant about how New Mexico handles their post-project evaluations of TMPs is that they maintain a database of post-project evaluations for future considerations.

The scope of post-construction evaluation is associated with individual projects or broader agency policies. At the project level, DOTs may prepare summative documents/reports for a project for application of lessons learned in future projects. At the agency level, the *Final Rule on Work Zone Safety and Mobility*, "requires agencies to perform a process review at least every two years to assess the effectiveness of their work zone safety and mobility procedures" (*1*). The rule gives three options for what can be reviewed: the DOT can evaluate work zone data at the agency level, the second option is to review randomly selected projects, and finally it is possible for the DOTs to do a combination of the first two options (*1*). Across the DOTs that were evaluated, it became apparent that most DOTs are compliant with this rule as they hold biennial (or more frequent) process reviews of either work zone data or randomly selected work zones throughout the states. However, there is great variability with how states conduct the reviews: some states have established one team to conduct these reviews while other states maintain multiple teams.

Examination of Procedures among Different State DOTs

This research examined the differences in how state DOTs approach project-level procedures, agency-level procedures, and types of data observed and collected during TMP implementation and post-construction evaluation.

Tier List Determination

Three tier lists were constructed for project-level procedures, agency-level procedures, and data observed during evaluation, each with three tiers (high tier, middle tier, and low tier) to distinguish exemplary state DOTs from their underdeveloped contemporaries. The aim of this categorization was to institutionalize more effective feedback procedures throughout the nation by identifying and expanding the methods practiced by the more successful DOTs.

In order to develop the three tier lists, DOT and agency manuals were reviewed for factors specifically pertaining to the TMP. Through a content analysis, key words and factors were observed and documented, and the states that practiced a certain factor would be noted. The factors that were noted involved the monitoring and feedback portions of the TMP. Table 2 has an overview of the factors. Following this, each category was assigned a weight representing its significance in feedback procedures. More weight was placed on factors that dealt with the evaluation of projects after the project had been completed than was placed on factors that monitored the project while it was in construction. The weighting also favored evaluation and monitoring that was performed by a transportation agency employee, rather than a contractor. Lastly, more weight was assigned to factors that happened more frequently. An example is the process reviews; annual process reviews were worth double biennial process reviews because it would allow for corrections to be made sooner. An arbitrary measure known as the effectiveness level was created. The following equation for evaluating effectiveness level was used. The effectiveness level can range from 0 to 100:

$$EL = \sum_{i=0}^{n} w_i * p_i \quad (1)$$

Table 2. Summary of Factors Investigated

Agency level processes	Project level processes	Data observed
• Frequency agency reviews were conducted • Description of how the agency reviewed projects (i.e., by jurisdiction or the state as a whole) • Review team made up of representatives from different divisions • If a feedback loop mechanism was present • Methods used to conduct agency-wide review	• Specified which personnel was responsible for the TMP during construction • Work zone reports • Feedback loop mechanism for individual projects (i.e., report or meeting)	• Work zone crash reviews and/or analysis • Peak traffic monitoring measures • Considerations taken when inspecting the work zone • The work zone was checked for compliance with TMP • Community input was utilized

Table 3. Tier List for Agency-Level Procedures

Tier	State DOTs
Top tier	Alaska, California, Colorado, Connecticut, Illinois, Kansas, Kentucky, Missouri, New Mexico
Middle tier	Arizona, Arkansas, District of Columbia, Delaware, Indiana, Iowa, Minnesota, New Hampshire, North Dakota, South Carolina, Vermont, Virginia
Bottom tier	Georgia, Idaho, Michigan, Mississippi, Montana, North Carolina, Oklahoma, Tennessee, Utah, Wisconsin

where EL = effectiveness level; w_i = weight assigned for category; and p_i = presence factor (0 for absence, 1 for presence).

Following observation of all factors, tier lists were devised based on the effectiveness levels DOTs had garnered. DOTs representing the top 25% were placed in the high tier, while the bottom 25% were placed in the low tier. The remaining DOTs were placed in the middle tier. The tier lists do not have the same states. If, when the state's transportation manual was reviewed, it did not contain any factors for a specific tier list, (i.e., agency-level processes) the state was not listed in the tier list. For example, Rhode Island is listed in the project-level processes table and data observed table, but it was omitted from the agency level processes table because no information about the agency reviews could be found in the manual that was reviewed.

Agency-Level Evaluation–Process Reviews

At the agency level, process reviews were thoroughly investigated: their frequency (annual or biennial), use of projects (randomly selected across the state or stratified by district/region), specific feedback processes, and methods and tools used at the process level (site visits and surveys of relevant personnel) were examined. Table 3 shows the tier list for agency-level procedures.

Two states representing the top tier regarding process reviews are Colorado and Missouri. Colorado brings together a team of five or six individuals (*10*). Colorado reviews at least six projects from each region, one of which is a nighttime operation if there is one ongoing. These reviews last around a week per region (*10*). The review team will look at a myriad of performance measures and other factors, such as geographic location and type of work to give the work zone a quality rating (*10*). The team then discusses the results at debriefing meetings, then they prepare two reports (*10*). Missouri's DOT has a "work zone review team" that is formed from personnel across the department (*18*). Missouri's review team conducts an inspection of all active work zones and reviews the strengths and weaknesses of the TMP processes of each district. The review team completes a report annually (*18*). Missouri also conducts daily quality assurance inspections of work zones done by the central office and district staffs (*18*). Missouri holds a work zone summit every year that covers new policies and initiatives for the next year (*18*). With all of the inspections and reviews that are conducted at the process level, Missouri allows for more data to inform revisions to their current TMP practices.

In comparison, Minnesota has a more average review process. Minnesota has four groups that work to provide improvement in the area of work zone safety and mobility (*19*). Both Missouri and Minnesota's review teams meet annually and review about three districts a year (*18, 19*). Minnesota has process feedback discussions that are based on work zone data or the results of field reviews

Table 4. Tier List for Project-Level Procedures

Tier	State DOTs
Top tier	Alaska, Delaware, Kentucky, Nevada, New Hampshire, New Mexico, North Carolina, South Carolina, Wisconsin
Middle tier	Arizona, California, Colorado, District of Columbia, Illinois, Maryland, Michigan, Missouri, North Dakota, Rhode Island, Utah, Vermont, Virginia
Bottom tier	Arkansas, Georgia, Indiana, Minnesota, Montana, New Jersey, Oregon, Tennessee

Table 5. Tier List for Data Observed during Evaluation

Tier	State DOTs
Top tier	Colorado, District of Columbia, Illinois, Kentucky, Michigan, South Carolina, Utah, Wisconsin
Middle tier	Alaska, Arizona, Arkansas, California, Connecticut, Delaware, Georgia, Indiana, Minnesota, Missouri, Montana, Nevada, New Hampshire, New Jersey, North Dakota, Vermont, Virginia
Bottom tier	Idaho, Iowa, Kansas, Maryland, Mississippi, New Mexico, North Carolina, Oregon, Rhode Island

that are held to recognize the best practices. The findings are then presented to a team sponsor in an annual review (*19*). Minnesota does an adequate job of reviewing their practices statewide, but there is room for improvement so that more data could be used in refining the TMP process.

Project-Level Evaluation

At the project level, personnel involved in TMP implementation (specifically, types of engineers, contractors, project leaders/coordinators, unspecified "responsible people," and management or planning teams were looked for), the application of work zone logs/diaries for continual observation during implementation, and feedback methodologies for projects (informative written summations/reports and post-project meetings or conferences) were observed. Data tables constructed from observing these factors led to the Project-Level Procedures tier list (Table 4).

An example of a state DOT with exemplary project-level evaluation procedures is Nevada. During construction, the resident engineer assigns a technician the duties of inspection and monitoring (including observation of traffic flow, flagging, the effectiveness of TTC plans, etc.) and the technician is responsible for daily reports back. Following the completion of construction, the project manager prepares an evaluation report for the TMP documenting lessons learned and suggests recommendations for improving TMP processes/guidelines. Post-construction reviews are conducted on all projects, and their purpose is to modify procedures to improve future projects. A quarterly report is produced and distributed by the Construction Division, identifying strengths and weaknesses of the TMP. However, some states are vastly underdeveloped in project-level evaluation; as an example, Montana simply states that someone is responsible for monitoring the TMP.

Depth of Data Observed during Evaluation

For data observed during evaluation, safety impacts (particularly feedback procedures regarding work zone crashes), mobility impacts (queue data, delay data, and traffic volume), factors related to field inspections (observing traffic control measures, consideration of time of day and geographic location, detours/closures, construction/maintenance signing, and pavement markings/conditions), quality assurance reviews, and community perception/input were taken into account. This yielded the Data Observed during Evaluation table (Table 5).

The tier list above shows the states that are the most insightful during their evaluations, taking a significant number of factors into consideration. Colorado DOT is particularly strong in its evaluation materials as it assesses the geographic location and pavement markings along with performance measures such as queue and delay data (*10*). Wisconsin and Washington, D.C. also have well-structured data points observed during their evaluation processes, describing the usefulness of the TMP and evaluating performance measures and the public reaction to the TMP (*16*, *17*).

Best Practices for Feedback Loop Procedures

By observing evaluation procedures at both project and agency levels and various factors considered by high-performing DOTs, the following include and describe some of the practices that effectively allow for continual improvement and pursuit of work zone safety and mobility:

- Daily work zone logs/diaries
  - Individuals responsible for TMP implementation should log their progress on a daily basis. Many DOTs use these for post-construction evaluation at both project and process level evaluation.
- Conferences/meetings for promotion of work zone safety and mobility
  - This applies both during and after construction. Many DOTs establish meetings to discuss necessary improvements: Colorado and Washington, D.C. do monthly meetings to determine necessary changes to the TMP plan during construction. Colorado and Missouri also do annual meetings with a broader overview, often observing a conglomeration of projects.
- Work zone crash reviews: Short-term feedback loop mechanisms
  - Crashes that occur in the work zone are evaluated by all state DOTs to establish safer conditions in future projects. Performing an evaluation for each individual crash and preparing a subsequent report within a set number of days (many DOTs set this at no later than ten days) allows for improvement of work zone safety and mobility during construction and in preparation for future projects.
- Reporting for each individual project
  - During the project, the responsible person should observe deficiencies in the TMP prepared during the design phase and report them to the agency or appropriate supervisor for change. Following construction, many DOTs prepare summaries of each project. Some DOTs compile these together and observe them collectively during broader agency-level evaluation activities, such as project reviews.
- Annual work zone reports: Crash data and traffic control devices
  - Some DOTs, such as the District of Columbia, prepare annual work zone crash reports and annual work zone traffic control inspection reports. These allow the agency to investigate these areas to adjust its policy, rather than restricting evaluation to individual projects.
- Process reviews: frequency
  - The Rule's requirement is to hold process reviews at least once every two years. Many DOTs have adjusted to making process reviews annual, allowing for more frequent feedback at the agency level.
- Process reviews: Scope of projects investigated
  - Stratifying projects observed during process reviews by district or region (some DOTs ensure that a specific quota is satisfied for each region) allows the conducting team to observe distinct factors within each district.
- Prioritizing community input and feedback
  - Attentiveness and consideration towards suggestions and criticisms from the public are vital towards establishing an effective feedback loop system; some DOTs even consider this to be at the forefront of evaluation. Developing methods of garnering such opinions, such as interviews with road users or an online feedback method, can give insight towards improvements for future projects.

Figure 1 shows an ideal TMP review process that a state DOT can implement by taking best practices identified in this study.

Conclusion

The goal of this research was to evaluate current practices of DOTs in TMP design, implementation, and post-

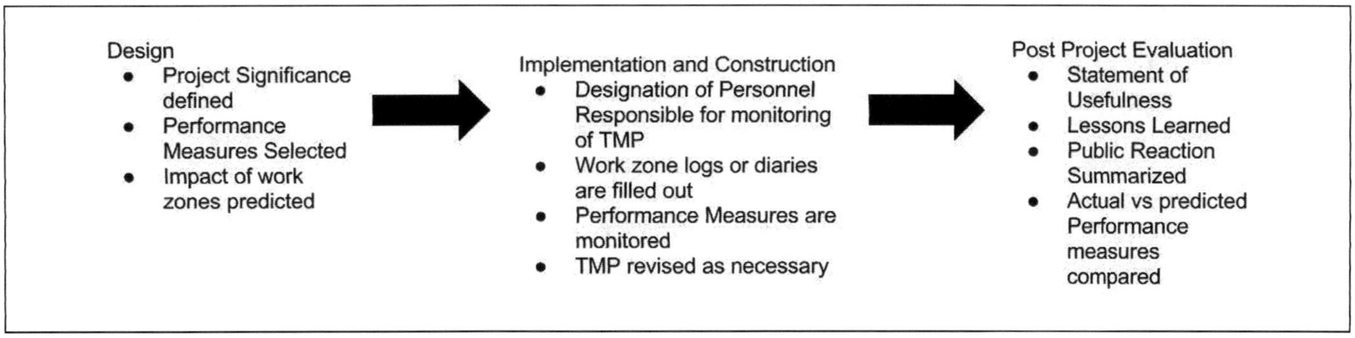

Figure 1. Ideal review process.

construction evaluation processes and identify the best practices which provide a benchmark for all DOTs to strive to meet as they seek to improve their individual evaluation procedures. This study provides a ranking of the different DOTs with regard to their TMP policies and procedures at the agency level, project level, and data used for TMP evaluation process. Some high-level conclusions from this study are that it is important to have evaluation policies in place during implementation and construction and the most effective post-project evaluations may need to a) have a statement regarding the usefulness of the TMP, b) consider the public reaction to the TMP, and c) compare the actual performance measures against the predicted performance measures. Process reviews are best when a multidisciplinary team is formed, and the team meets annually to review the work zone processes and all available work zone data. The process review is also helped by inspecting the work zones that are active during the review. As a continuation of this study, the research team will conduct structured interviews with top-tier DOTs to get a better picture of the current practices and efforts being made by DOTs to improve their procedures that might not be in writing yet. This information can be used to later develop a maturity assessment and guidance for advancement.

Author Contributions

The authors confirm contribution to the paper as follows: study conception and design: MV, RJ, JS; data collection: MV, RJ, JS; analysis and interpretation of results: MV, RJ, JS; draft manuscript preparation: MV, RJ, JS. All authors reviewed the results and approved the final version of the manuscript.

References

1. FHWA. Work Zone Safety and Mobility Rule (23 CFR 630 Subpart J). In *Final Rule on Work Zone Safety and Mobility – FHWA Work Zone*. Federal Highway Administration, n.d.Accessed July 27, 2017.
2. Gallo, A. A., L. E. Dougald, and M. J. Demetsky. *Development of Performance Assessment Guidelines for Virginia's Work Zone Transportation Management Plans*. Report No. FHWA/VCTIR 13-R6. Virginia Department of Transportation, Richmond, 2012.
3. South Carolina Department of Transportation. *Rule on Work Zone Safety and Mobility*. N.p., 2009.
4. Valerio, M. Infrastructure Design Directive IDD-2009-02 (Work Zone Traffic Control). 2009.
5. California Department of Transportation. *Transportation Management Plan Guidelines*. Publication N.p., 2015.
6. Delaware Department of Transportation. *Delaware DOT Work Zone Safety and Mobility Procedures and Guidelines (Chapter V, VI) (2007)*. N.p., 2007.
7. Tennessee Department of Transportation. *Work Zone Safety and Mobility Manual*. N.p., 2017.
8. Oregon Department of Transportation. *ODOT Three Levels of TMPs – Sample Transportation Management Plans – FHWA Work Zone*. Federal Highway Administration, 2017.
9. Nevada Department of Transportation. *Nevada DOT Work Zone Safety and Mobility Implementation Guide (4,5.4))*. N.p., 2015.
10. Colorado Department of Transportation. *CDOT Work Zone Safety and Mobility Rule Procedures Document*. N.p., 2014.
11. Alaska Department of Transportation and Public Facilities. Design and Construction: Highway Work Zone Safety and Mobility. *Policy and Procedure Manual*, Alaska Department of Transportation and Public Facilities, Juneau, AK, 2007. http://www.dot.alaska.gov/admsvc/pnp/assets/chapt_5/05_05_015.pdf.
12. Bourne, J. S., E. Chung, G. L. Ullman, D. Gomez, B. Zimmerman, T. A. Scriba, R. D. Lipps, D. L. Markkow, K. C. Mathews, D. L. Holstein, and R. Stargell. Best Practices in Work Zone Assessment, Data Collection, and Performance Evaluation. *Scan 08-04*. Arora and Associates, P.C., Lawrenceville, NJ, 2010. https://ops.fhwa.dot.gov/wz/docs/nchrp_20_68a/nchrp_20_68a.pdf.
13. Georgia Department of Transportation. *Work Zone Safety and Mobility Policy*. N.p., N.d.
14. Illinois Department of Transportation. Chapter Thirteen Work Zone Transportation Management Plans. In *Bureau of Design and Environment Manual*. N.p., 2010.
15. *2016 Work Zone Safety Grant*, Vol. I. Technical Application Institute for Transportation, Iowa State University, 2016.
16. Wisconsin Department of Transportation. Chapter 11 Section 50 Design Traffic Control. In *Facilities Development Manual*. N.p., N.d.
17. Washington D.C. Department of Transportation. *District of Columbia Work Zone Safety and Mobility Policy*. N.p., 2007.
18. Missouri Department of Transportation. Work Zone Safety and Mobility Policy. *Engineering Policy Guide*. Missouri Department of Transportation, Jefferson City, MO, 2017. http://epg.modot.org/index.php/616.14_Work_Zone_Safety_and_Mobility_Policy.

The Standing Committee on Construction Management (AFH10) peer-reviewed this paper (18-04804).

Article

TRR
JOURNAL OF THE TRANSPORTATION RESEARCH BOARD

Transportation Research Record
2018, Vol. 2672(26) 114–123
© National Academy of Sciences:
Transportation Research Board 2018
Article reuse guidelines:
sagepub.com/journals-permissions
DOI: 10.1177/0361198118776133
journals.sagepub.com/home/trr
SAGE

Feasibility of using QR Codes in Highway Construction Document Management

Hyun Woo Lee[1], Bharat Anand Harapanahalli[2], Chukwuma Nnaji[3], Jonghyeob Kim[1], and John Gambatese[3]

Abstract

Highway construction occasionally takes place in remote locations, making its document management challenging especially when frequent document revisions occur. With the recent advancement of smartphones and tablets, Quick Response (QR) codes can provide project teams rapid and reliable access to up-to-date documents required for field operations. As a result, the use of QR codes can lead to a reduced need for traveling or meeting for document revisions, and reduce the amount of hardcopy documents and storage space. Despite the potential for significant benefits, there have been few studies aimed at assessing the feasibility of using QR codes in highway construction. In response, the objective of the study was to investigate the benefits of and barriers to using QR codes in highway construction for document management. To conduct the study, first a multi-step process was used, involving an online survey and interviews, with a goal of determining the status quo of highway construction in terms of document management and mobile information technology (IT). The results indicate that hardcopy documentation is still the most prevalent form of document management in highway construction, and hence there is an opportunity for implementing QR codes in conjunction with mobile IT. In the second part of the study, a time study using a real-world infrastructure project was conducted based on three activities: detail look up, specification check, and version check. As a result, the study found statistical evidence that using QR codes can lead to significant time savings.

Government agencies such as state Departments of Transportation (DOTs) invest a considerable amount of capital in infrastructure projects, especially in highways and bridges. According to the American Society of Civil Engineers (ASCE) Report Card (*1*), the Federal Highway Administration (FHWA) invests approximately $12.8 billion to repair and maintain bridges yearly and it anticipated an additional $8 billion to be needed to meet the future demand. These infrastructure projects help to stimulate economic growth and define the regional economic conditions (*2*, *3*). Therefore, the timely completion of these projects is important, yet the projects are subject to delays that can result in additional construction costs (*4*). These delays are associated with various factors such as topographical challenges, design errors, untimely document disseminations, and constantly changing surrounding conditions during construction (*5*, *6*).

Document Management and Use of IT in Construction

Effective communication and sharing of information has always been a critical component of a project's success (*7*). Currently, the construction industry is exploring the use of information technology (IT) to improve communication methods for the successful completion of projects. Moreover, companies are currently using computer software and electronic documents extensively for managing the flow of information in their projects (*8*). According to Kim et al. (*9*), one of the important uses of IT in construction is to electronically manage project documents. This combination can increase the efficiency and effectiveness of managing documents electronically, especially on the Internet and cloud servers (*10*). Further, Stewart et al. (*11*) stressed employing an appropriate framework to increase the acceptance rate of IT in construction. In recent years, there have been studies in the field of IT and document managing systems involving the development of frameworks to facilitate IT in the construction industry. One such study by Yunna and Na

[1]Department of Construction Management, University of Washington, Seattle, WA
[2]Hallmark Constructions, Malleshwaram, Bangalore, India
[3]School of Civil and Construction Engineering, Oregon State University, Corvallis, OR

Corresponding Author:
Address correspondence to Hyun Woo Lee: hyunwlee@uw.edu

(*12*) developed a framework enabling the management of documents in the initial phase of a project. Further, Zhu et al. (*13*) recognized the prevalence of hardcopy documents in the industry and developed a framework to include such documents in an electronic database. In addition, an in-depth case study was conducted by Hjelt and Bjork (*14*) to identify factors that influence the adoption of IT and electronic document management systems (EDMS) by construction companies. This research eventually developed into a framework to incorporate IT and information management systems in organizations. Moreover, in the past few years, online EDMS based on cloud servers have provided a convenient platform for project teams to collaborate and communicate effectively (*10*, *15*). These studies show that the use of online EDMS in the industry has been increasing significantly in recent years.

Smartphones and tablets have been widely used in the construction industry in recent years. This trend has led various researchers to develop frameworks to incorporate mobile IT in construction communication. For example, Zhu et al. (*16*) developed a project-specific framework to support safety management procedures for scaffolding using mobile IT. Zhiliang et al. (*17*) developed a framework to incorporate mobile IT for efficient material management in the field. Chen and Kamara (*18*) developed a framework to facilitate mobile IT application to manage information in the field. These examples show that mobile IT can be effectively applied to various construction operations.

Based on the findings from the previous studies, integrating mobile IT and wireless networks into document management can support project stakeholders in highway construction. These systems allow stakeholders to better coordinate and solve document-related issues that may arise during projects. Moreover, quick and reliable access to these systems can avoid unnecessary time spent in travelling and meetings.

QR Codes

Past studies have suggested that combining the use of barcodes with administrative systems such as document management systems can increase system reliability and ease information access (*19*). The use of barcodes has proven to be one of the most reliable tools to access and retrieve information in the field. Barcodes are represented in the form of unique black and white patterns, which—when scanned by a laser or digital camera—allow rapid access to the information stored in them (*19*). Barcodes have been extensively used in the retail and manufacturing industries to keep track of inventories. Barcodes are used as an error-free input of data into the systems, making the data entry and transfer reliable (*20*, *21*). However, one-dimensional barcodes started being replaced with two-dimensional barcodes that offer higher storage capacity and higher responsiveness (*20*). According to Lin et al. (*22*, *23*), two-dimensional codes are cheaper, easy to produce, more secure, and capture a variety of data types.

Quick Response (QR) codes are the most widely used two-dimensional barcodes that consist of several unique black and white pixels and are square in shape. The high capacity of QR codes makes them suitable to store data of larger size (*24*). The commercial use of QR codes has led to increased awareness amongst public users. In particular, the availability of smartphones and tablets has enabled users to access the data encoded in QR codes using applications, providing access to the digital world from the physical world (*25*).

QR Codes in the Construction Industry

In the past few decades, barcodes have been successfully used in the construction industry for improving the accessibility of information for field personnel and for the management of inventories in the field. The use of barcodes has supported the industry involving a tremendous amount of data that has to be coordinated and shared among project members.

According to McCullouch and Lueprasert (*20*), QR codes can be applied to identification cards, maintenance management, shipping processes, processing equipment repair, and on material safety data sheets. The various existing applications of QR codes in the industry indicate the potential for its use in document management.

There have been previous efforts to use QR codes in construction document management mainly for version checking. A system called isOK^{TM} attempted to use QR codes to enable quick access to revised documents (*26*). The system proved to be effective in reducing the amount of rework by providing access to the latest version of the drawings (*27*). It is important that field personnel have easy access to project documents and that they are immediately notified of any changes to avoid errors and further complication in the process. These needs provide an opportunity to implement QR codes combined with mobile devices.

Point of Departure

Although multiple attempts have been made by previous researchers to implement mobile IT and EDMS in the building sector, there have been few studies aimed at identifying the current document management practices or applying mobile IT and EDMS in highway construction. In response, the objectives of this study are to investigate the current document management practices in highway construction and to propose QR codes as a medium for the industry to effectively improve its document management using mobile IT and EDMS. The study results are expected to provide an opportunity for the highway construction industry to adopt advanced IT for the real time sharing of project information in the field.

Methodology

An investigation via a survey and semi-structured interviews was conducted to determine the current practice of using mobile IT in the highway industry and to assess its application to construction document management in the field. Including a survey and interviews provides means to concurrently triangulate contextual information that supports and informs the need for QR codes in highway construction document management. In addition, the survey and interviews were intended to provide an understanding of the industry's awareness of using QR codes and mobile IT for managing documents in a project.

Following the collection and analysis of contextual data, the researchers developed and tested the implementation of QR codes for highway construction document management. To achieve this, an embedding protocol was designed and implemented for a time study that was conducted to determine the potential benefit that QR codes can offer for highway construction document management. Specifically, the time study was conducted for detail look up, specification check, and version check. Below is a synopsis of activities performed in each key method.

Survey

The survey questions mainly aimed at identifying:

1. The existing document management methods or procedure currently used by the survey participants and their organizations,
2. The extent to which mobile IT has been incorporated in the highway industry, and
3. The knowledge levels for mobile IT and QR codes.

The survey was hosted and administrated via Google Drive™ for approximately 50 days and was mainly distributed to highway construction professionals within the Associated General Contractors of America (AGC) Oregon Columbia Chapter and industry partners of Oregon State University (OSU).

Interviews

The interviews aimed to elicit responses similar to the three main questions asked in the survey, albeit with more detailed information. Open-ended questions were designed and administered in person, over the phone, or via email for highway contractors within Oregon and Washington.

QR Code Time Study

In preparation for the time study, QR codes were embedded in the plans of an Oregon DOT construction project which led the participants to an EDMS online database that provided various pieces of project information. The EDMS online database and its webpages were established using Google Sites™ for the time study. Access to the database during the study was enabled by Wi-Fi provided by OSU.

Results and Discussion

Survey

In total, 26 complete responses were received from the sampled population. The survey results revealed that the participants were currently working in California, Idaho, Nevada, Oregon, and Washington. Their roles in projects range from project managers (34.6%), project engineers (23.1%), estimators (23.1%), field engineers (7.7%), to an area manager (3.8%), and an executive (3.8%). Further, the participants were involved in different types of highway construction projects, including new construction (36.96%), reconstruction (23.91%), resurfacing (15.22%), and rehabilitation and restoration (15.22%). The remaining 8.7% of the responses were from participants who work on non-highway projects. It was also found that 77% of the participants indicated 5 years or more of professional experience.

The survey participants indicated that they visit the field multiple times every day and typically carry hardcopies of plans, specifications, estimates, and other construction documents. In short, the survey found that field personnel in highway construction still heavily rely on hardcopy documents for sharing and retrieving data. Consequently, the survey results imply that using mobile IT for document management in highway construction has the potential to benefit the industry.

When asked about how document revisions are disseminated to the field, the participants indicated that emails (50%) are the most preferred method for document revision notifications, followed by meetings (18.75%) and phone calls (16.67%). Large projects tend to use software that provides automatic electronic notifications to field personnel. However, documents are still distributed manually in the field, as mentioned by one participant.

Next, when asked about the frequency of revision notifications, 30.8% of the participants said they are notified multiple times a week, followed by at least once every other week (23.1%), once a month (23.1%), and once a quarter (3.8%). Such a high frequency of revisions emphasizes that quick access to the documents can be beneficial for field personnel. Lastly, when asked about the time taken to disseminate revised documents, 35% of the participants said within a day, whereas 15% were notified of document revisions within two weeks from the day of the revisions.

In terms of IT proficiency, 88.5% of the participants indicated that they are proficient with smartphones or tablets for conducting email conversations and viewing plans and specifications. Furthermore, it was found that the participants with 11 to 15 years of professional experience utilize their mobile devices relatively more than other groups. This finding indicates that the use of mobile devices is widely accepted by highway construction professionals irrespective of their experience in the industry.

When asked about the barriers to using mobile devices on site, the participants selected fast battery consumption (26.09%), small size of the screen (41.30%), poor network reception (26.09%), and difficulty in operating the device (6.52%). When further asked about the purpose of using their mobile devices in the field, it was found that 52% of the participants use it mainly for viewing plans and specifications, whereas 84% use it for browsing information on the Internet. Therefore, the survey shows that a significant number of the sampled field personnel currently use mobile devices as part of their daily activities in the field. In addition, 96% of the participants use the devices specifically to check emails, and this can be connected to the current document management practice that mostly relies on emails for revision notifications. In contrast, the reasons participants reported for not using mobile devices in the field (11.5% of the participants) were, that mobile devices were not necessary for their scope of work, and their lack of knowledge of how to use mobile devices.

When asked about the knowledge level regarding QR codes and their application, the survey results revealed that 19.2% of the participants did not have any knowledge about QR codes, whereas 23% know what QR codes are but were unaware of their application. Nonetheless, 39% have used or scanned QR codes at least once. The lack of knowledge or awareness of QR codes in the industry seems to be a significant barrier. In addition, 16.7% of the participants still expressed their hesitation in adopting such a new method in their projects. However, it is important to note that 45.8% of the participants support the implementation of QR codes for document management and provided positive feedback about their contribution to the industry. A few participants mentioned that the use of QR codes can benefit the industry as they provide field personnel with rapid and reliable access to up-to-date documents using mobile devices. One participant emphasized that the opportunity to access information from a project database would help to manage and transfer important data without having any data loss during the process.

Interview

Five interviews were conducted with four highway contractors in the states of Oregon and Washington. Two interviews were conducted by phone, two in person, and one via email.

Due to the recent advancement of IT and increased use of mobile devices, three contractors reported that they have partially implemented mobile IT in their document management process. In other words, they have not expanded their use of mobile IT to a full scale EDMS. In contrast, one contractor reported that their use of mobile IT in the field is limited to making phone calls and checking emails only. Additionally, two other contractors employ EDMS which use web-based services such as Dropbox™ and Google Drive™. Irrespective of the contractors' use of EDMS, all the interviewees reported that they still carry a significant amount of hardcopy drawings to the field, and each set can consist of 1,000 pages on average. In addition to the drawings, a set of specifications is also carried to the field.

The interviews revealed that the size and frequency of document revisions in a project vary depending on the complexity of the project. An interviewee reported that at the time of the interview, his project had just received the 99th revision package and, on the other hand, another contractor said they only had two design changes during their paving project. Further, it was observed from the interviews that the dissemination process of revised drawings varies from project to project. However, all interviewees currently do not employ a well-defined framework that systematically enables field personnel and subcontractors to access revised documents in real time. Field personnel and subcontractors are typically notified of revised documents either through phone calls, meetings, or emails.

An interviewee from a project that employs an EDMS responded that the project's superintendent and project manager can view the revised drawings on their mobile devices. However, depending on their personal preference or in the presence of significant changes, a hardcopy of the revised drawing is printed for reviewing the changes. As a summary, Figure 1 depicts the three steps of the current document management system in highway construction, based on the findings from the interviews.

Time Study

QR Code Embedding Protocol. A QR code can generally be developed in two different ways: using websites that

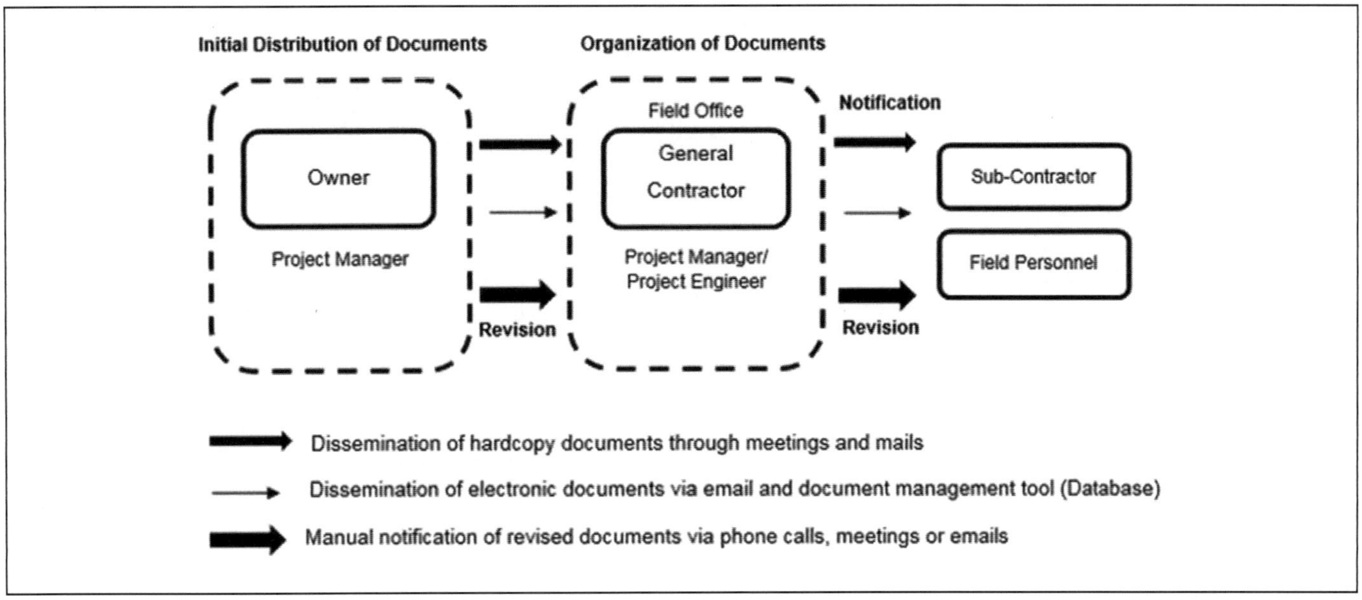

Figure 1. Current document management process (*28*).

allow generating QR codes or using AutoCAD's QR plug-ins. First, some websites allow for quick generation of QR codes. In doing so, one must be careful about which data source (e.g., web address) has to be converted. The websites typically require the web addresses of source documents to be entered in a specific space. Then, an image of the QR code is generated in one click. The image must be stored by the user for its further use—embedding in target documents. Another way to generate a QR code is to use AutoCAD. The QR plug-in provided by AutoCAD enables quickly converting any type of text into QR codes (*29*). The plug-in is included in AutoCAD and codes can be generated free of charge.

Linking a QR code is one of the crucial steps in the implementation of QR codes. The QR code image saved by the user can be embedded into drawings or any type of document. One must ensure that the image size of the QR code is large enough for mobile devices to scan. Between the two ways of generating QR codes, using the plug-in in AutoCAD was found to be more convenient because codes and drawings can be developed simultaneously. Figure 2 shows an example of embedded QR codes in a 2D drawing, which was used in the time study.

Time Study Structure. The time study was conducted at OSU under the researchers' supervision in a controlled environment (e.g., a room free of distractions). The study was performed using Nexus 10 (a tablet computer co-developed by Google and Samsung Electronics, running the Android operating systems) with a QR code application.

The database created for the study was intended to mimic a web-based EDMS like the one used by contractors, which hosts electronic versions of project documents including drawings and specifications. The database was structured in such a way that each individual webpage supports a particular specification or standard drawing ("secondary document" in Figure 3) that is tied to a specific part of the bidding plan ("primary document" in Figure 3). These documents were formatted as PDFs making it easier for the participants to view and manage the files with their mobile devices. Therefore, by using the database, QR codes, and the tablet device, the study was structured to replicate real-life scenarios that field personnel may perform in the field on a daily basis.

Adapting the study of Yeh et al. (*30*), the time study involved three activities: (*1*) detail look up, (*2*) specification check, and (*3*) version check. The bidding plans, specifications, and addenda of an Oregon DOT project were used for each of the three activities, in their respective order. The study participants were asked to find and report specific information that was listed in a questionnaire at the commencement of the study. Each of the three activities comprised of three cases, Cases A, B, and C, each requiring the participant to refer to and report on information from the primary documents (e.g., site plan in Case A) and subsequently locate a reference element from the corresponding secondary documents (e.g., site utility detail in Case A). Most importantly, the time taken for each process was measured—time taken from when exploring the primary documents commenced to when the required information was located in the secondary documents. Figure 3 illustrates the process.

The main objective of the time study was to compare the time taken by two methods, namely, with-QR-codes

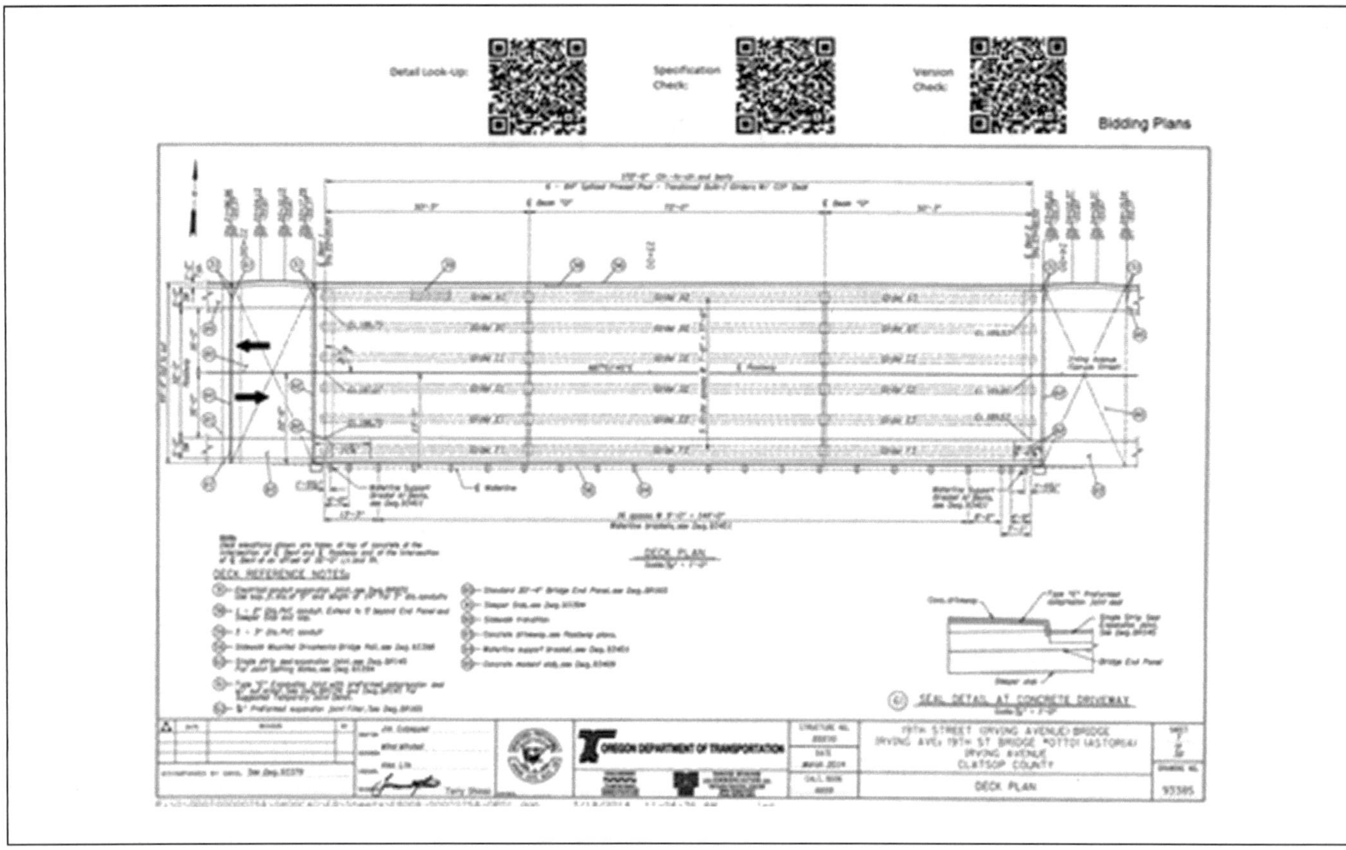

Figure 2. QR codes embedded drawing used for time study (*28*).

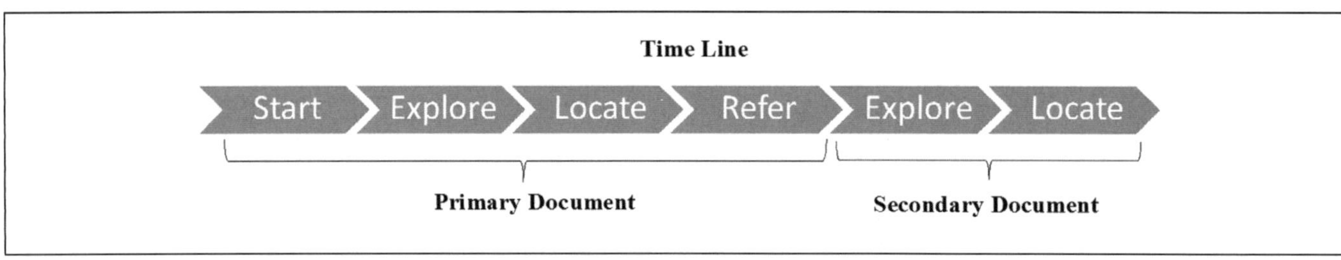

Figure 3. Time study process (adapted from *28*).

versus without-QR-codes. The without-QR-code method followed the steps typical of the industry to extract information from documents. This process required a participant to answer the first question from the questionnaire by exploring and locating the answer from the primary documents and then to answer the second question by exploring and locating the answer from the secondary documents. For the with-QR-code method, identical activities were performed until the first question was answered (Step 1), and then embedded QR codes were scanned (Step2) to answer the second question by using the tablet to access the EDMS's online database (Step 3). Figure 4 depicts the process flow for the with-QR-code method.

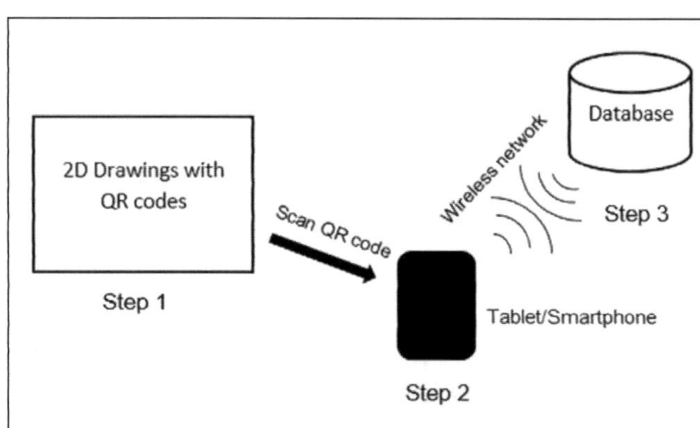

Figure 4. Process flow for with-QR-code method (*28*).

Time Study Results. Twenty-five students in the School of Civil and Construction Engineering (CCE) at OSU participated in the time study. The students were randomly selected from various levels of classes offered in CCE and most were majoring in Construction Engineering Management. At the start of the time study, a preliminary survey consisting of a set of preliminary questions was performed to identify their construction experience, awareness of QR codes, proficiency with mobile devices, and proficiency with drawing reading (hereinafter, preliminary survey). The efficiencies were measured on a Likert scale of 1 to 5 (1 = poor; 2 = somewhat poor; 3 = neutral; 4 = good; 5 = very good).

The results of the preliminary survey showed that 64% of the participants had 6 months or more of construction experience through part-time jobs or internships. They indicated different levels of experience in using drawings and specifications of a highway construction project. For example, 24% of participants indicated that they possess "good" proficiency in reading construction drawings based on their prior construction experience and through courses that involved drawing reading. Further, 84% of the participants said that they were familiar with the use of QR codes. The participants with limited experience in plan reading were provided with additional instructions on how to access the required information by using a different set of heavy civil drawings that were not part of the test.

Upon the completion of the preliminary survey, the participants were briefed about the steps that they would follow to complete the activities required for the time study. In addition, a brief demonstration was given about how to use the tablet that was provided to the participants, and then the participants were given an opportunity to scan a QR code to make sure they understood how to use the tablet.

On completion of the time study, the collected time study data was analyzed using a nonparametric Wilcoxon sign test to compare the differences in time taken between the activities that were performed with and without QR codes for each case (A, B, and C). The analysis was performed using IBM's SPSS statistical software.

The results of the Wilcoxon test indicate that the activities for detail lookup and specification check can be performed more efficiently using QR codes than without QR codes (p-value < 0.05 for all cases) under ideal conditions. It is believed that the without-QR-code method took significantly more time because of the extra time consumed in exploring the required page or information in the secondary documents, which was performed after referring to the primary documents. In other words, the entire process was simplified by employing the QR code method, in which the QR codes quickly directed the participants to the relevant webpage, where the secondary documents related to the primary documents were easily found. However, the time difference was insignificant for the version check (p-value > 0.05). This may be attributed to the limitation of the time study in replicating a real-life situation for checking the versions of documents.

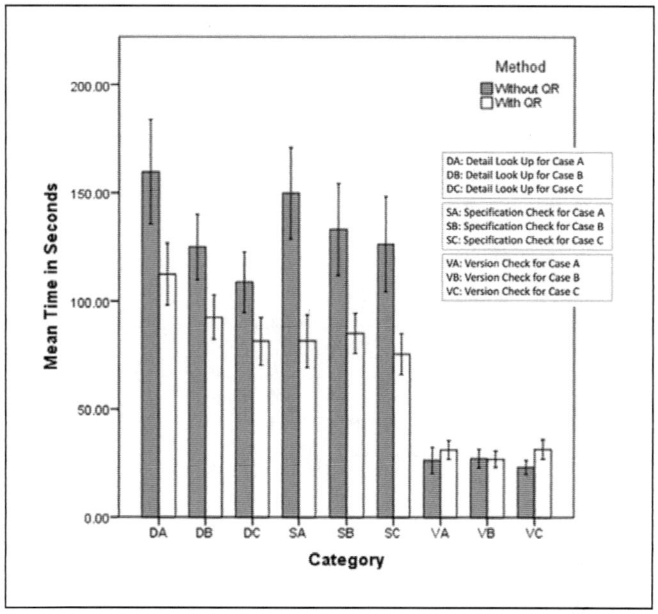

Figure 5. Mean time for each category (time in seconds) (*28*).

Figure 5 presents a bar chart representing the mean time taken in seconds for the three activities. It can be seen from the error bars that there exist significant differences between all the cases of detail lookup and specification check. In other words, it is visually noticeable that the with-QR-code method resulted in significantly less time than the without-QR-code method in the cases of detail lookup and specification check, but not for version check. D, S, and, V in Figure 5 stand for detail lookup, specification check, and version check, respectively, for each case. For example, DA refers to detail look up for Case A.

Figure 6 shows box plots that were created to examine the spread and skewness of observations for each category. It is seen that the with-QR-code method resulted in not only shorter time but also more consistency in time than the without-QR-code method, confirming the benefit of using QR codes for document management. It should be noted that the outliers as seen in Figure 6 are due to external issues such as slow Internet speed or improper handling or functioning of the tablet during the time study, which can also be seen in real-life.

In addition to the statistical analysis of the results of the time study itself, nonparametric Spearman's rank-order correlation was used to analyze the relationships

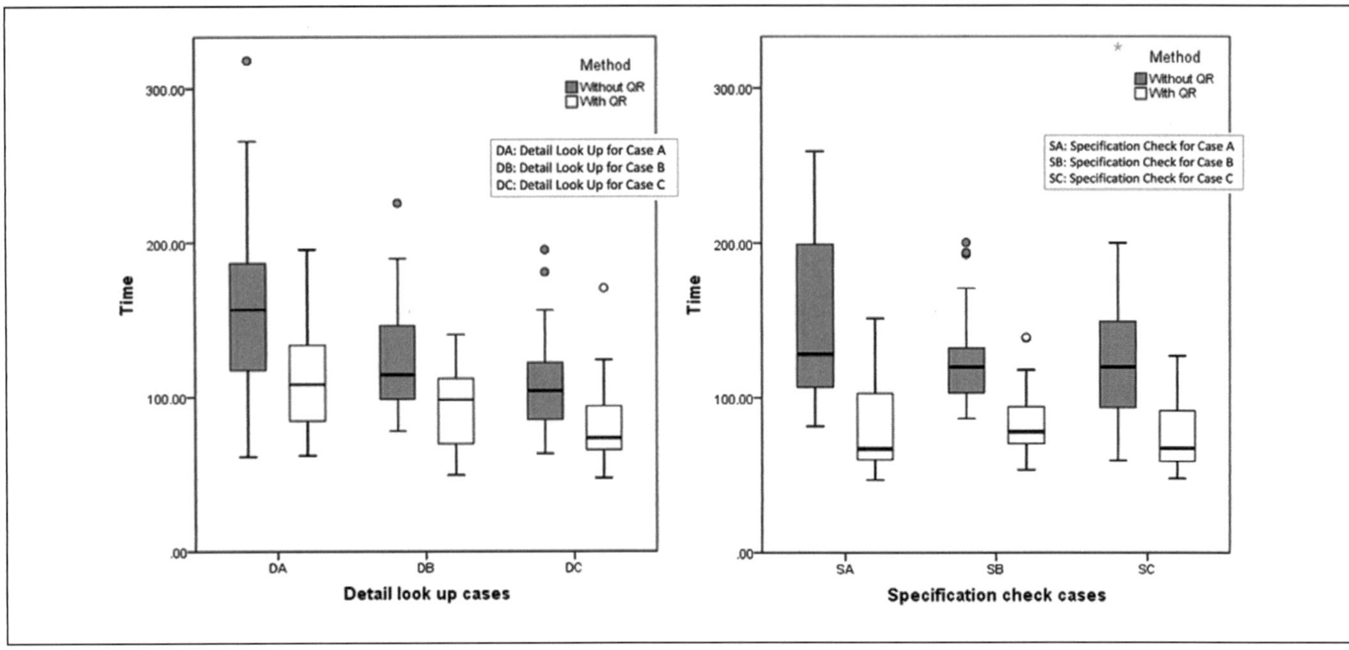

Figure 6. Results of time study in box plots (time in seconds) (*28*).

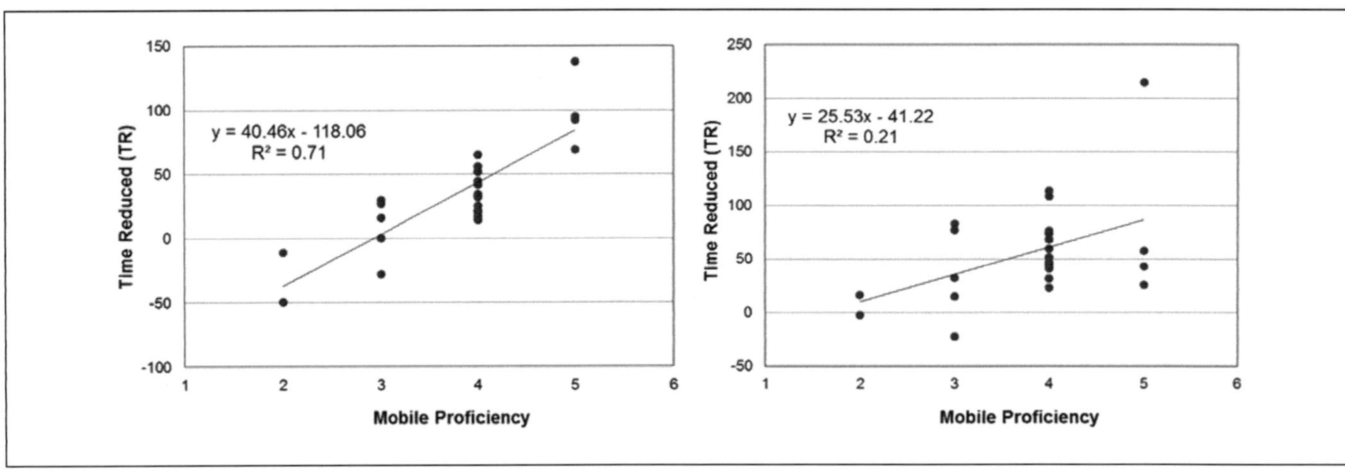

Figure 7. Mobile proficiency versus detail lookup TR and specification check TR (*28*).

between the responses in the preliminary survey with the results of the time study. As explained previously, during the preliminary survey, the participants were asked about their construction experience, mobile IT experience, awareness of QR codes, mobile proficiency, and drawing reading proficiency. For simplification purposes, the average time values of the three cases (A, B, and C) were calculated for each activity and the difference of these averages (hereinafter, Time Reduced (TR)) between the with-QR-code and without-QR-code methods for a particular activity were used for the analysis.

The result of the nonparametric correlation analysis indicates a significant, positive relationship (a correlation coefficient of 0.81) between the mobile proficiency and the TR of detail lookup (p-value < 0.01). In addition, it was also found that mobile proficiency is positively correlated with the TR of specification check. Its correlation coefficient was found to be 0.52 (p-value < 0.01). In short, statistical evidence was found that a participant with higher mobile proficiency indicates an increased TR of detail lookup and specification check. Figure 7 shows scatter plots of the mobile proficiency versus the TR of detail lookup and specification check, respectively. The mobile proficiency shown on the x-axis is based on the Likert scale as mentioned previously (1 = very poor to 5 = very good) and the y-axis represents the TR in seconds. Consistent with Figure 6, Figure 7 shows that some outliers were present in the study. Nevertheless, the analysis methods implemented are considered sufficient since both nonparametric methods are less sensitive to outliers or heavy tails.

The relationship between the mobile proficiency and the difference in the TR of detail lookup—which can represent time efficiency—can be quantified by using Equation 1:

$$\text{Time Reduction } (D) = 40.46 \times \text{Mobile Proficiency} - 118.06 \tag{1}$$

where D refers to detail look-up. The R-squared value is 0.71, implying that 71% of the variance in the TR is explained by the mobile proficiency. Similarly, the relationship between the mobile proficiency and the TR of specification check can be quantified by using Equation 2:

$$\text{Time Reduction } (S) = 25.53 \times \text{Mobile Proficiency} + 41.22 \tag{2}$$

where S refers to specification check. An R-squared value of 0.21 implies that 21% of the variance in the TR can be attributed to the mobile proficiency.

Regardless of the potential benefits postulated by the time study, the time study revealed that using QR codes may face barriers, including participants' lack of familiarity with smartphones and tablets and Internet service interruption. It should be noted that using a different test subject demography that includes construction professionals could impact the test result. Nevertheless, the results from the study demonstrate that QR codes can effectively be integrated into highway construction document management systems.

More information on the QR code embedding process, time study structure, and benefit–cost analysis results can be found in the PacTrans research report via the following link: (http://depts.washington.edu/pactrans/wp-content/uploads/2013/11/PacTrans-37-OSU-Lee.pdf)

Conclusions and Future Research

The present research study set out to evaluate the viability of utilizing QR codes as a tool for enhancing document management of highway construction projects, thereby filling a gap in literature. The study findings indicate that hardcopy versions of plans and specifications remain the prevalent means of accessing documents on highway projects. In addition, although most highway construction workers have access to mobile technology and the Internet, there remains some resistance to using EDMS. Results from the time study concluded that the application of QR codes to access electronic documents is feasible, and can provide field personnel with access to documents stored in EDMS. This research also offers multiple opportunities for future research as follows:

First, the time study data did not indicate statistical evidence that the use of QR codes is beneficial for version check. Therefore, future research is suggested to develop a structure to mimic real-life situations for version check. Second, the effectiveness of QR codes could be further validated by testing the implementation of QR codes in a pilot project. This validation is imperative given that the time study was conducted using students in ideal conditions that provided a relatively high-speed Internet connection, organized hardcopy documents, and absence of hazardous construction activities. Therefore, the pilot testing can identify additional benefits and barriers based on what has been identified in this research. In addition, a benefit–cost analysis should be performed to determine the financial viability of using QR codes in highway construction projects. Lastly, future study is suggested to investigate the types of information that can be encoded in QR codes depending on the requirements of projects.

Acknowledgments

This research was funded in part by PacTrans (Grant #DTRT12-G-UTC10).

Author Contributions

The authors confirm contribution to the paper as follows: study conception and design: Lee, Harapanahalli; data collection: Harapanahalli, Nnaji; analysis and interpretation of results: Lee, Harapanahalli, Nnaji, Gambatese; draft manuscript preparation: Lee, Nnaji, Kim, Gambatese. All authors reviewed the results and approved the final version of the manuscript.

References

1. American Society of Civil Engineers. *2013 American Society of Civil Engineers Report Card for America's Infrastructure*. American Society of Civil Engineers, Reston, VA, 2013. https://www.infrastructurereportcard.org/a-big-wiin-for-water-resources/. Accessed May 15, 2017.
2. Chandra, D., and E. Thompson. Does Public Infrastructure Affect Economic Activity? Evidence from the Rural Interstate Highway System. *Regional Science and Urban Economics*, Vol. 30, No. 4, 2000, pp. 457–490.
3. Khan, S., and D. Levy. Linking Economic Development to Highway Improvements: Pine Ridge Reservation, South Dakota. *Transportation Research Record: Journal of the Transportation Research Board*, 2003. 1848: 106–113.
4. Anastasopoulos, P., S. Labi, A. Bharagava, and F. Mannering. Empirical Assessment of the Likelihood and Duration of Highway Project Time Delays. *Journal of Construction Engineering and Management*, Vol. 138, No. 3, 2012, pp. 390–398.

5. Finch, E. F., R. Flanagan, and L. E. Marsh. Electronic Document Management in Construction Using Auto-ID. *Automation in Construction*, Vol. 5, No. 4, 1996, pp. 313–321.
6. Ellis, R. D., and H. R. Thomas. The Root Causes of Delays in Highway Construction. Presented at 82nd Annual Meeting of the Transportation Research Board, Washington, D.C., 2003.
7. Molenaar, K., and D. Navarro. Key Performance Indicators in Highway Design and Construction. *Transportation Research Record: Journal of the Transportation Research Board*, 2011. 2228: 51–58.
8. Al Qady, M., and A. Kandil. Document Management in Construction: Practices and Opinions. *Journal of Construction Engineering and Management*, Vol. 139, No. 10, 2013, 06013002.
9. Kim, S., Y. Bai, Y. Jung, and D. Yun. Highway Construction Productivity Measurement with a Wireless Real-Time Productivity Measurement System. *Transportation Research Record: Journal of the Transportation Research Board*, 2011. 2228: 26–33.
10. Björk, B. C. Document Management–a Key IT Technology for the Construction Industry. *Proc., the 2nd Worldwide ECCE Symposium Information and Communications Technology in the Practice of Building and Civil Engineering*, Espoo, Finland, Association of Finnish Civil Engineers, Helsinki, Finland, 2001, pp. 35–43.
11. Stewart, R. A., S. Mohamed, and R. Daet. Strategic Implementation of IT/IS Projects in Construction: A Case Study. *Automation in Construction*, Vol. 11, No. 6, 2002, pp. 681–694.
12. Wu, Y., and Y. Na. Framework System Design of Construction Project Document Management During the Initial Phase Based on WBS. *Proc., International Conference on Wireless Communications, Networking and Mobile Computing (WiCom 2007)*, Shanghai, China, IEEE, New York, NY, 2007, pp. 5248–5251.
13. Zhu, Y., W. Mao, and I. Ahmad. Capturing Implicit Structures in Unstructured Content of Construction Documents. *Journal of Computing in Civil Engineering*, Vol. 21, No. 3, 2000, pp. 220–227.
14. Hjelt, M., and B. Björk. End-user Attitudes Toward EDM Use in Construction Project Work: Case Study. *Journal of Computing in Civil Engineering*, Vol. 21, No. 4, 2007, pp. 289–300.
15. Alshawi, M., and B. Ingirige. Web-enabled Project Management: An Emerging Paradigm in Construction. *Automation in Construction*, Vol. 12, No. 4, 2003, pp. 349–364.
16. Zhu, X., X. Xue, M. Chen, and H. Zhou. Mobile IT Used in Construction: A Scaffolding Safety Management. *Proc., ICCREM 2013*, Karlsruhe, Germany, American Society of Civil Engineers, Reston, VA, 2013, pp. 482–491.
17. Ma, Z., Z. Liu, and D. Zhang. An Integrated Mobile Material Management System for Construction Sites. *Proc., AEI 2013*, State College, Pennsylvania, Building Solutions for Architectural Engineering, Reston, VA, 2013, pp. 354–363.
18. Chen, Y., and J. M. Kamara. Using Mobile Computing for Construction Site Information Management. *Engineering, Construction and Architectural Management*, Vol. 15, No. 1, 2008, pp. 7–20.
19. Shehab, T., and O. Moselhi. An Automated Barcode System for Tracking and Control of Engineering Deliverables. *Proc., 2005 Construction Research Congress, American Society of Civil Engineers*, Reston, VA, 2005, pp. 1–9.
20. McCullouch, B. G., and K. Lueprasert. 2D Bar-code Applications in Construction. *Journal of Construction Engineering and Management*, Vol. 120, No. 4, 1994, pp. 739–752.
21. Blakey, L. H. Bar Codes: Prescription for Precision, Performance, and Productivity. *Journal of Construction Engineering and Management*, Vol. 116, No. 3, 1990, pp. 468–479.
22. Lin, Y. C., W. F. Cheung, and F. C. Siao. Developing Mobile 2D Barcode/RFID-based Maintenance Management System. *Automation in Construction*, Vol. 37, 2014, pp. 110–121.
23. Lin, Y. C., Y. C. Su, and Y. P. Chen. Developing Mobile BIM/2D Barcode-based Automated Facility Management System. *The Scientific World Journal*, Vol. 2014, 2014, 374735.
24. Tarjan, L., I. Šenk, R. Kovač, S. Horvat, G. Ostojić, and S. Stankovski. Automatic Identification Based on 2D Barcodes. *Proc., XV International Scientific Conference on Industrial Systems (IS'11)*, Novi Sad, Serbia, University of Novi Sad, 2011, pp. 130.
25. Friedman, J. *QR Codes: Beyond the Trend.* Massaro Corporation, Pittsburgh, PA, 2012. http://massarocorporation.com/2012/09/qr-codes-beyond-the-trend/. Accessed March 15, 2017.
26. Dougherty, P. *isOK™ Revision Control System. isOK™ Patent Pending Construction Document Revision Control System*, Nashville, TN, 2011. http://promo.isok.us/isok-constructiondocument-revision-control-system/. Accessed March 15, 2017.
27. NCE Editorial. Costain/Laing O'Roourke Discover Cunning Use for QR Codes in Construction. *Online News, New Civil Engineer*, London, UK, 2012. http://www.nce.co.uk/news/bim/costain/laing-orourke-discover-cunning-use-for-qrcodes-in-construction/8627885.article. Accessed May 19, 2017.
28. Lee, H. W., and B. Harapanahalli. Evaluating the Feasibility of using QR Codes in Highway Construction Document Management. Pacific Northwest Transportation Consortium (PacTrans), Seattle, WA, 2015. http://depts.washington.edu/pactrans/wp-content/uploads/2013/11/PacTrans-37-OSU-Lee.pdf. Accessed May 11, 2017.
29. Autodesk Inc. *QR Codes, Apps Exchange*, Autodesk Inc., San Rafael, CA, 2014. http://apps.exchange.autodesk.com/ACD/en/Detail/Index?id=appstore.exchange.autodesk.com%3aqrcodes%3aen. Accessed May 10, 2017.
30. Yeh, K. C., M. H. Tsai, and S. C. Kang. On-site Building Information Retrieval by Using Projection-Based Augmented Reality. *Journal of Computing in Civil Engineering*, Vol. 26, No. 3, 2012, pp. 342–355.

The Standing Committee on Construction Management (AFH10) peer-reviewed this paper (18-00941).

Any opinions, findings, conclusions, or recommendations expressed in this report are those of the authors and do not necessarily reflect the views of PacTrans.

Article

TRR
JOURNAL OF THE TRANSPORTATION RESEARCH BOARD

Transportation Research Record
2018, Vol. 2672(26) 124–133

DOI: 10.1177/0361198118773886
journals.sagepub.com/home/trr
SAGE

Effects of Reheating Procedure and Oven Type on Performance Testing Results of Asphalt Mixtures

Zachary Lemke[1], Husam Sadek[1], Daniel Swiertz[2], Signe Reichelt[3], and Hussain U. Bahia[1]

Abstract

Reheating and oven-aging procedures of plant-produced asphalt mixtures in laboratories are important topics to consider as performance testing of mixtures becomes more popular among agencies. Differences between laboratory equipment and procedure could significantly affect performance properties. The objective of this study is to investigate the influence of sample size, oven type, and variation in reheating/aging temperatures on the results of two performance tests on plant-produced mixtures. A selected mixture was tested for volumetric properties and performance using Hamburg wheel-tracking (HWT) and semi-circular bending (SCB-IFIT) tests. Results show that reheating mixtures uncovered and in smaller containers could significantly reduce the time to achieve aging temperature, and could make the process more efficient and consistent. In addition, aging using three different oven types showed that temperature within ovens can vary significantly depending on the location of the sample inside the oven, which affects the time required to reach the target temperature, and thus may also influence the aging of the sample. The mixture volumetric properties show that the effect of various heating conditions is marginal. Using the developed reheating/aging procedure of this study, the results of the HWT and SCB-IFIT tests showed no substantial effect of oven type on rutting and cracking resistance. The overall results indicate that there is a need to standardize the conditions of reheating, sample geometry, and to verify uniformity of temperature in ovens. Such standardization can further reduce variability and thus should be part of the AASHTO/ASTM standard procedures for quality control, or of laboratory equipment calibration procedures.

Hot mix asphalt (HMA) undergoes significant physical and mechanical changes during plant production and during service in a process collectively known in the paving industry as aging. Laboratory protocols simulating these processes are an ongoing topic of research. Two high-profile NCHRP studies are centered on developing laboratory procedures to simulate aging of asphalt mixtures during production and during service.

Short-term aging predictive procedures were developed as the primary objective of the NCHRP 09-52 study (*1*) for aging of asphalt mixtures for mix design and performance testing to simulate the effects of plant mixing and processing. Long-term aging procedures are the focus of the ongoing NCHRP 09-54 study (*2*) to be calibrated and validated with field data to simulate field aging of mixtures for performance testing and prediction. Both studies focus on the procedure of conditioning and aging mixtures to simulate plant and field; however, the effect of the sample size, oven type, and variation in reheating/aging temperatures and durations on the aging level and the results of performance tests are not discussed.

Recent research, however, has investigated the effects of laboratory conditions and procedures (such as equipment type, frequency of opening oven doors, ambient sample storage, time, etc.) on the mechanical and volumetric properties of the aged mixtures (*3*). The effects of laboratory practice, such as repeated opening and closing of oven doors, has also been discussed recently during Asphalt Mixture Expert Task Group Meetings.

One overlooked aspect of creating laboratory aging procedures is a method to reheat cooled mixture specimens delivered from plant production. It is expected that the amount of sample and geometry would affect reheating times, but to what extent has not been studied. Considering the current trend for increased use of binder modifiers, recycled materials, and warm mix asphalt technologies, a consistent procedure needs to be established for reheating and

[1]Department of Civil and Environmental Engineering, University of Wisconsin-Madison, Madison, WI
[2]Bitumix Solutions, a Div. of H.G. Meigs, LLC, Portage, WI
[3]Behnke Materials Engineering, LLC, Beloit, WI

Corresponding Author:
Address correspondence to Zachary Lemke: zlemke@wisc.edu

aging asphalt mixtures to achieve representative results in the laboratory that simulates field conditions.

In this study, a standardized reheating and aging procedure was developed based on several preliminary studies to assess the effect of sample size and variation in reheating/aging temperatures on the duration and efficiency of the aging process. Then, the effect of oven type on the results of performance tests for a selected asphalt mixture was investigated. The mixtures were subjected to two main performance tests: Hamburg wheel-tracking (HWT) test for rutting resistance, and the semi-circular bending (SCB) test for cracking resistance. The findings of this study are intended to be used in support of an asphalt mixture aging study sponsored by the Wisconsin Department of Transportation, Wisconsin Highway Research Program in which plant-produced mixture specimens will be aged in the laboratory and tested for performance.

For the purposes of this study, all mixture conditioning and aging is conducted on loose mix samples, and no claim is made as to the representation of these methods to field conditions. This work is intended to identify factors in the laboratory (and between laboratories) that can affect the mixture aging process in an effort to standardize mixture-handling procedures.

Materials

In this study, one plant-produced mixture was used for all testing. The mixture utilized in this study was sampled and boxed by an independent contractor for use in an agency "round robin" mixture testing program. The mixture has a total binder content of 5.4% and is a Wisconsin Standard Specification "E-3" mixture, which is a design designated for roads with a 20-year design equivalent single axle load (ESAL) range of 1–3 million. The mixture utilized PG58S-28 asphalt binder and the binder replacement level for the mixture is approximately 26%. The mixture was completely cooled before beginning any testing.

Selection of Aging & Conditioning Procedures

Current guidance on short-term oven aging (STOA) of asphalt mixtures for intended use in mechanical property testing from AASHTO R30 is conditioning of uncovered, loose mixture specimens for four hours at 275°F, stirring every hour. Long-term oven aging (LTOA), per the same AASHTO specification, is conducted on compacted specimens for 120 hours at 185°F with a cooling period of about 16 hours after aging.

Several research studies have concluded that the AASHTO-recommended STOA protocol may actually be longer than what is needed to simulate plant production of mixtures, and a more reasonable duration is 2 hours of loose mix aging at 275°F, stirring after 1 hour (*4*, *5*). There appears to be consensus in the literature that loose mix aging best simulates short-term aging during plant production.

There is no consensus, however, regarding the aging conditions for LTOA. Experience in Wisconsin on pilot projects has shown the practical difficulty in following AASHTO guidelines when performance testing of mixtures is needed for acceptance in the field. This led to the implementation of loose mixture aging at 275°F for 12 hours for these projects. Findings for Wisconsin mixtures suggest that this protocol "produced comparable asphalt binder properties to compacted mix aging under the current AASHTO R30 protocol," although no claim as to the representation of this procedure to field mixtures was made (*4*). Recent work conducted in support of NCHRP 09-54 has found that loose mix aging above 212°F may "alter oxidation mechanisms," however (*6*).

Oven Types

For the oven comparison study, three separate ovens (in three different laboratories) were used for reheating and aging of the loose mix samples: Oven-A, Oven-B, and Oven-C. Oven-A uses a high-volume fan in which the convicted heat is moved through stainless-steel ducts that are on each side of the oven. The intake of fresh air is fixed, as well as the exhaust rate, as it is regulated by a damper that is located on the back of the unit.

Oven-B operates in a similar fashion to Oven-A as the heating element is on the top and the hot air flows from the sides of the oven. The hot air originates in the back and moves forward as it circulates. The sides of the interior are made up of an aluminized steel. Finally, Oven-C utilizes a stainless-steel interior and circulates hot air from the back to the front and across the shelves. Figure 1 shows the hot air circulation inside the three ovens used in this study.

Development of Reheating and Aging Procedure

To develop a procedure for reheating and aging of asphalt mixtures in this study, several tests were conducted to evaluate the effect of sample size, oven type, and the location of the sample inside the oven. The following sections summarize the results of the tests and significant findings.

Duration to Reheat Boxed versus Unboxed Mixtures

In Wisconsin, it is standard procedure to sample plant-produced mixtures in cardboard boxes to aid in transport and storage for "round robin"-type testing; for this study, the boxes contained approximately 17 kg of mixture. After delivery of the boxed mixture to the laboratory for testing, the box is reheated to be remixed and representatively split for further testing. This initial step, namely the reheat

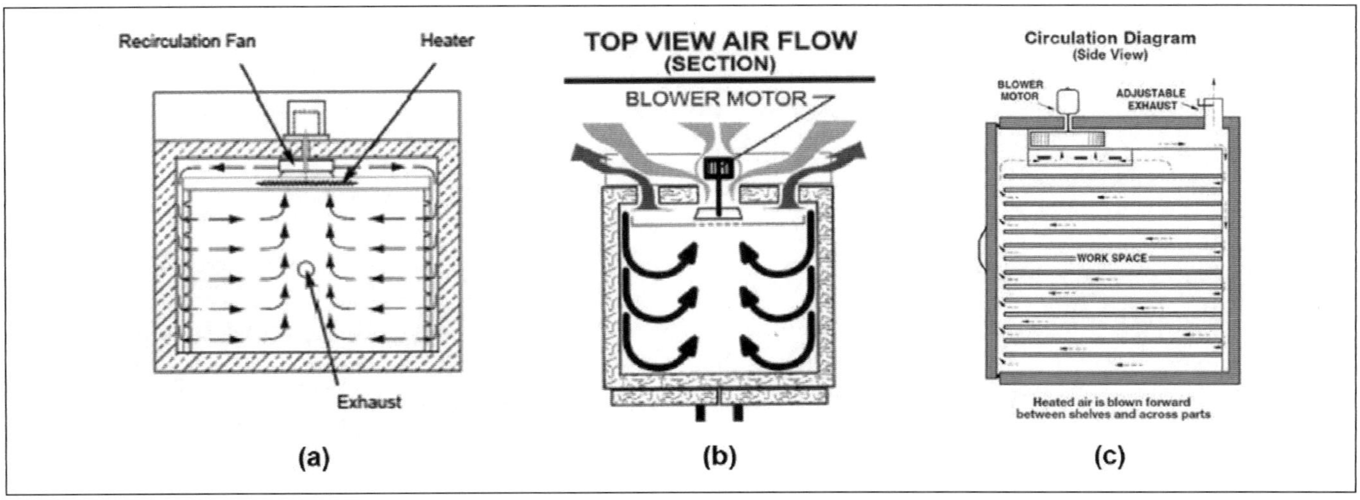

Figure 1. Air flow of: (*a*) Oven-A, (*b*) Oven-B, and (*c*) Oven-C.

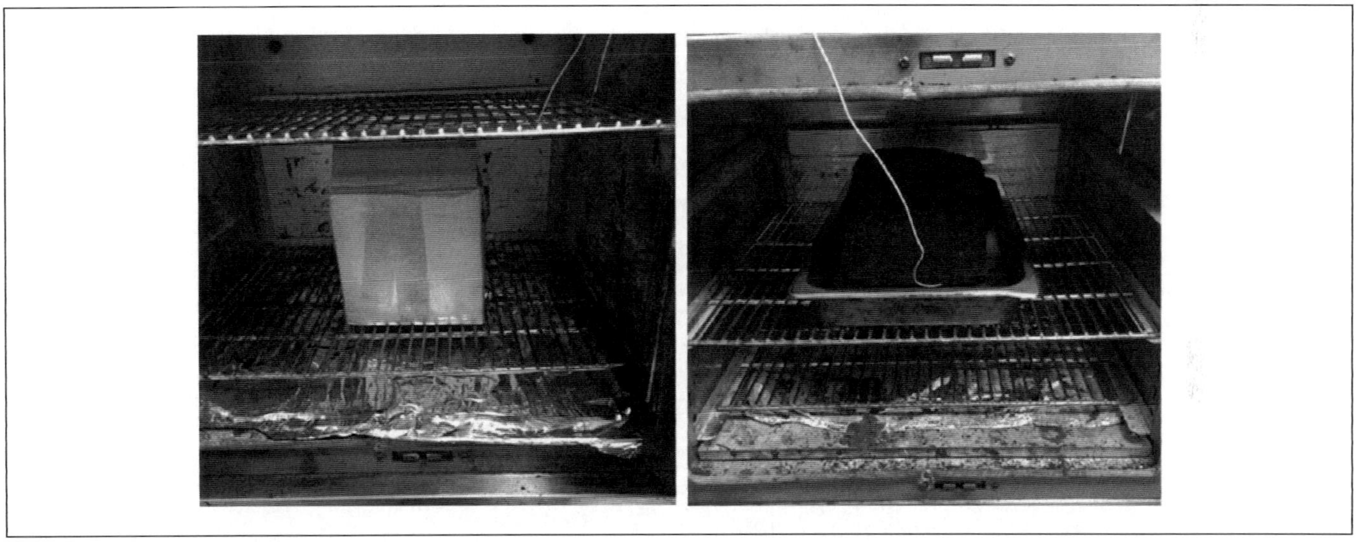

Figure 2. Reheating boxed versus unboxed HMA blocks.

temperatures and duration, has yet to be standardized in Wisconsin. Therefore, it was important to evaluate the time required to reheat boxed and unboxed mixtures.

One boxed sample and one unboxed sample (a boxed sample that had the cardboard removed and placed in a metal pan) were reheated to determine the duration it took for the block to reach a temperature at which it could be split and mixed representatively, as shown in Figure 2.

After 2 hours of heating, a temperature probe cast within the unboxed sample block reached 215°F (102°C) whereas the boxed sample only reached 135°F (57°C). Ultimately, the higher temperature of 215°F after 2 hours produced a mixture that could be broken apart and split representatively. This finding confirmed that there is a significant effect of unboxing the sample block on the time required for reheating.

Duration to Reheat Covered versus Uncovered Mixtures

On finding that the unboxed mixture reached the splitting temperature faster, the study progressed to investigate whether after splitting the sample into pans, covered or uncovered mixtures reach the target aging temperature quicker.

The reheated sample block was remixed and placed into two stainless-steel pans (11" × 9" × 2.5") with a thickness of 1.25 inch. One pan was covered with aluminum foil and the other was left uncovered in the middle rack of the oven preheated to the desired aging temperature of 275°F (135°C). Figure 3 shows that the uncovered sample took approximately 25% less time to reach the desired temperature (i.e., 275°F) relative to the covered sample. Again, the effect of covering the sample during

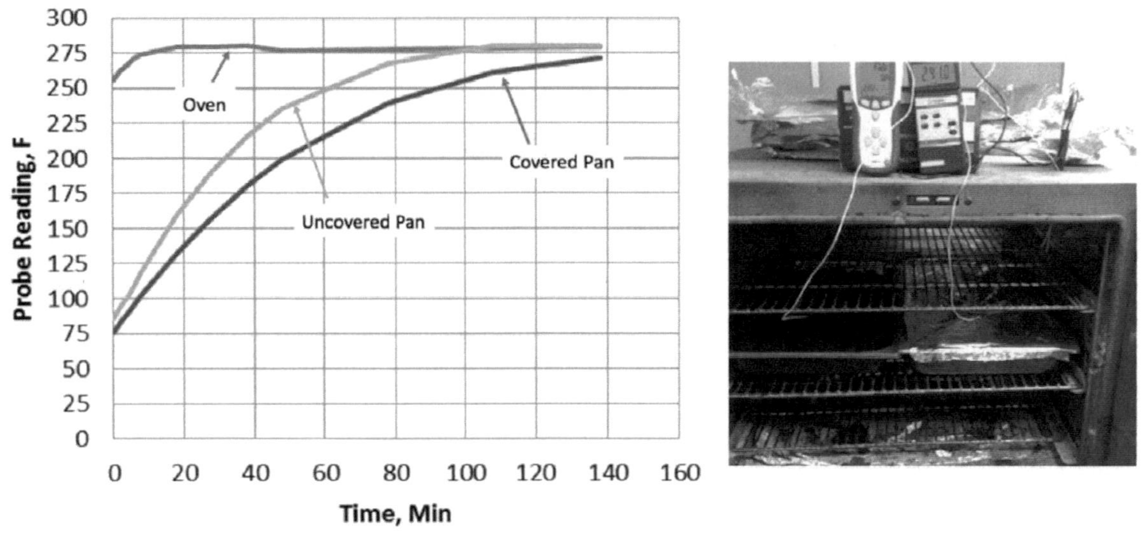

Figure 3. Monitoring temperature of covered versus uncovered samples.

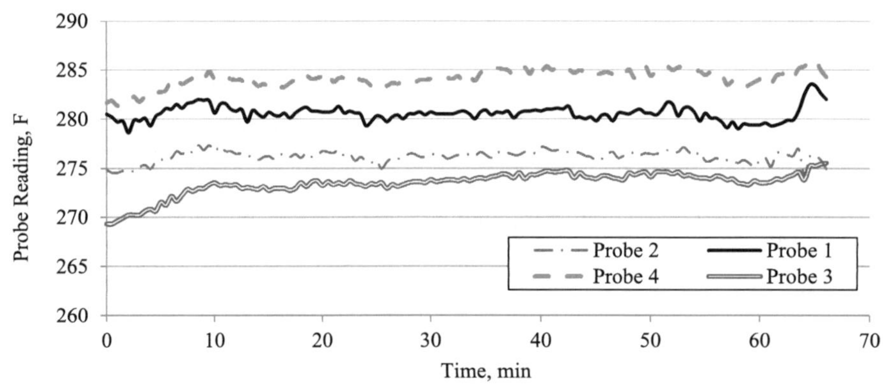

Figure 4. Temperature of probes at specified locations in the oven.

reheating and/or aging can slow down the reheat process significantly.

Location of Sample in Oven

Location of the mixture inside the oven during the reheating process may seem trivial, but the oven set point may not represent the temperature at all locations within the oven. Some locations within the oven may be warmer or cooler than others; therefore, calibration may be needed for all ovens before use. To check temperature uniformity within an oven in this study, four probes were placed in different locations in the same 275°F oven. Recorded temperatures showed that the temperature distribution is not uniform in all locations of this oven, as shown in Figure 4. Probe 4 in the bottom left corner of the oven was the hottest location with an average difference of about 15°F compared with Probe 3, the coolest location. The difference in probe readings for this oven varied by as much as approximately 10°F (AASHTO R30 has a tolerance window of 10.8°F for conditioning).

Furthermore, although the thermostat was reading correctly when measured against a traceable reference thermometer with a probe in the same location, at least one area of the oven was outside of current AASHTO R30 limits; this finding suggests that although the oven set point (and thermostat accuracy) can be checked, the actual mixture temperatures at the sample location should be monitored. These results constituted further data to be collected using Oven-A. Temperature distribution using Oven-A appeared to be more uniform. It was found that individual ovens, even from the same model, should be calibrated and temperature distribution can be affected by oven type.

To investigate the effect of location inside the oven on mixture reheating, four uncovered HMA samples in pans

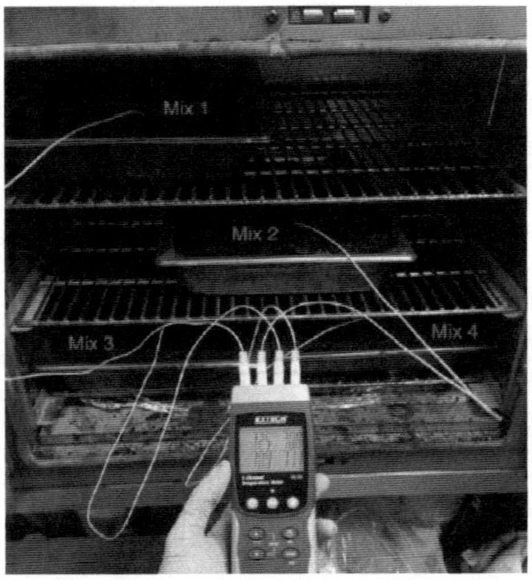

Mix-Location #	Sample Location	Time to AASHTO Lower Limit (269.9°F), min	COV, %
Mix-1	Left of top rack	193	23.0
Mix-2	Middle of center rack	135	
Mix-3	Left of bottom rack	115	
Mix-4	Right of bottom rack	139	

Figure 5. Four locations of uncovered mixtures in the oven.

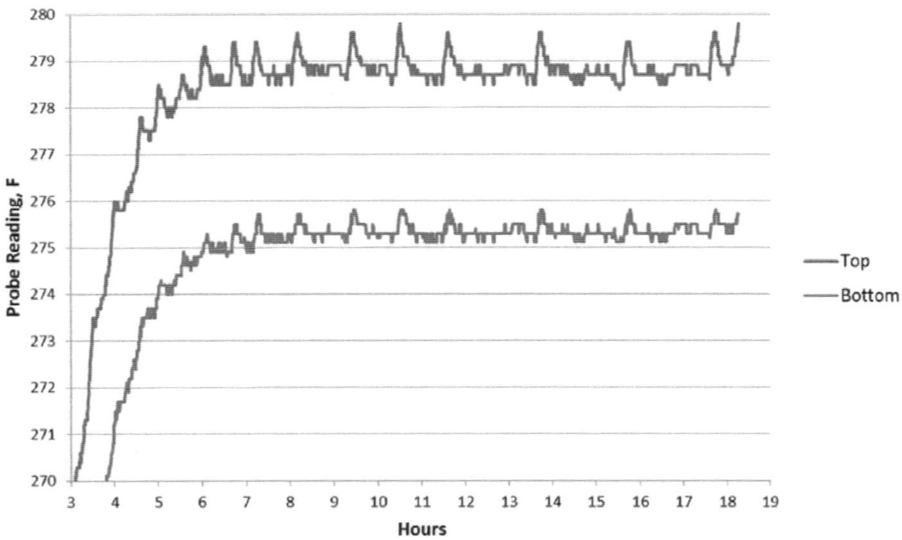

Figure 6. Temperature of two HMA samples undisturbed for 18 hours.

(1.25-inch mixture thickness) were placed in the 275°F (135°C) Oven-B at different locations, as shown in Figure 5.

All samples were at room temperature at the start of the test and the probes were cast 1 inch deep into the center of the pan. The table in Figure 5 outlines the duration it took each mixture, according to its location in the oven, to reach AASHTO R30 lower limit of oven-aging temperature (269.9°F).

As per the results shown, sample position in the oven is demonstrated to affect the reheat time significantly. In addition, temperature distribution is not necessarily uniform in the oven, particularly when it is fully loaded. This will be expanded upon in the next section.

Oven Loading

The effect of oven loading on the temperature distribution and the reheat duration was investigated in this study. Two mixture samples were placed in the upper-center and the lower-center of Oven-B. These samples were left undisturbed and the temperature was monitored for more than 18 hours, as recorded in Figure 6. The results show that the mixes reached AASHTO R30 lower limits after 3 hours and 3.6 hours for top and bottom locations, respectively. Therefore, when the oven is not fully loaded temperature distribution appears to be better compared with the previous section when the oven was fully loaded, with a slight difference in reheating time, with respect to location in the oven.

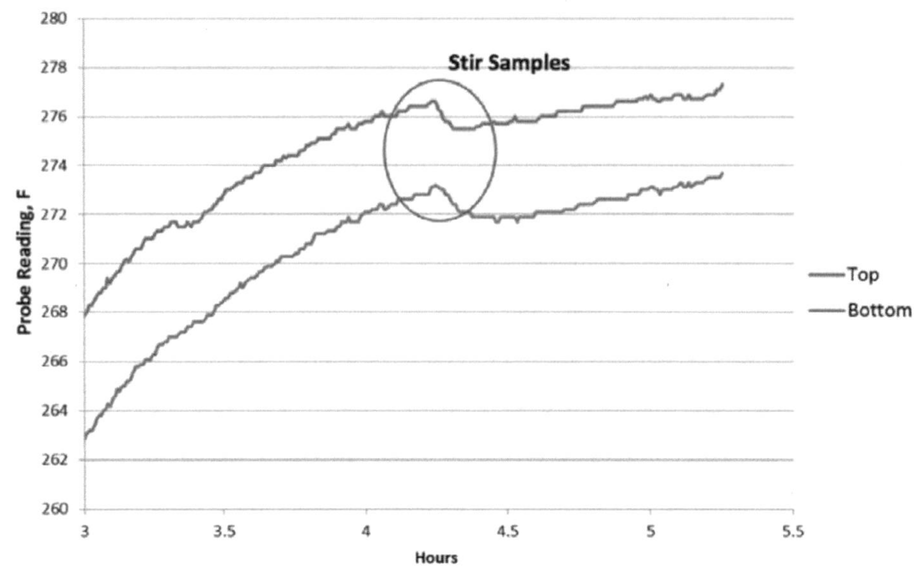

Figure 7. HMA samples' temperature at top and bottom rack of the oven before and after stirring.

Opening Oven to Stir Samples

Simulating the short-term aging of mixture samples requires stirring of the sample 1 hour after reaching the initial target temperature. In this part of the study, the effect of opening the oven door to stir samples on the mixture temperature was investigated.

Two HMA sample pans were uncovered and placed in upper-center and lower-center locations of Oven-B. These samples were stirred 1 hour after reaching the AASHTO R30 lower limit. Stirring the loose mixture in each pan was completed in under 1 minute. Figure 7 shows the temperature of the mixtures monitored before and after stirring.

As shown in the figure, the top mixture returned to pre-stir temperature in about 38 minutes whereas the bottom mixture took 51 minutes. However, neither mixture fell outside of AASHTO limits during the stirring process. From this study, it is concluded that although opening the oven to stir the mixture specimen will cause a reduction in mixture temperature, if the duration of stirring is kept to a minimum, the decrease in temperature may not be significant (using AASHTO R30 temperature ranges as a benchmark). Furthermore, the time required for the mixture to return to the pre-mixing temperature can vary significantly, which in turn may be an issue if the oven is opened repeatedly during the aging process.

Effect of Temperature Accuracy of Oven on Performance of Aged Mixes

Although the initial testing showed clear effects of sample covering and location in the oven, it was important to determine if variations in mixture temperatures can affect the results of volumetric properties and performance in terms of rutting resistance.

Table 1. Average Air Voids, Gmm, Rut Depth of the Samples Aged at 270°F and 285°F

Parameters	Samples aged at 270°F	Samples aged at 285°F
Average air voids, %	7.1%	6.6%
Average Gmm	2.526	2.528
Average creep slope, mm/pass	−0.43	−0.48
Average rut depth at 20,000 passes, mm	3.9	4.1

A total of six mixture samples were prepared in pans for reheating and aging at two temperatures, 270°F and 285°F. Each aging temperature had three replicate samples, one for determining the theoretical maximum specific gravity (Gmm) value and two for the HWT test. The Gmm sample pan for each aging temperature was reheated and aged as the only sample in the oven on the middle-center rack. The two HWT pans for each temperature were placed in the oven at the same time, side by side on the middle-center rack. All six samples were STOA and followed the same reheating and aging procedure. The samples were uncovered during the reheating and aging sequence. The Gmm, the bulk specific gravity (Gmb), and HWT procedures were tested as per ASTM D6857 (*7*), AASHTO T331 (*8*), and AASHTO T324 (*9*), respectively.

Table 1 shows the variance among air voids, Gmm, and the HWT test results between the two aging temperatures. The results clearly show that the oven temperature affected only the air void content (however, the air voids were still within the AASHTO T324 tolerance, 7% ± 1%, for the test),

Table 2. Gmm Results for Each Sample per Oven Type

Aging term	Sample ID	Gmm	Average		D2s		Standard deviation
STOA	Oven-A-1	Damaged	2.481	2.484	–	0.003	0.003
	Oven-A-2	2.481					
	Oven-C-1	2.485	2.487		0.003		
	Oven-C-2	2.488					
	Oven-B-1	2.482	2.483		0.001		
	Oven-B-2	2.483					
LTOA	Oven-A-1	2.515	2.499	2.497	0.033	0.013	0.013
	Oven-A-2	2.482					
	Oven-C-1	2.501	2.498		0.007		
	Oven-C-2	2.494					
	Oven-B-1	2.483	2.494		0.022		
	Oven-B-2	2.505					

while the Gmm and the HWT results showed no effect of the temperature change. Therefore, it can be concluded that the accuracy of the oven temperature during reheating and/or aging, even when slightly affecting the percentage of air voids, does not affect the performance of this mixture tested using the HWT test significantly.

Results of Oven Type Comparison Study

The oven comparison study utilized the same three ovens previously mentioned and began with two mixture blocks. Each mixture block was removed from the cardboard boxes, heated at 300°F (149°C) for 2 hours and then remixed and separated into seven flat aluminum pans (3,700 grams per pan) for the STOA and an 18-hour LTOA processes; 18 hours was chosen to represent an extreme example for this study, and not necessarily to correlate to field aging. The seven pans were left uncovered and cooled to room temperature before the aging process. Each type of oven aged a group of seven pans that were then tested in one laboratory to determine the Gmm for STOA and LTOA samples, the rutting resistance for STOA samples using the HWT test, and the cracking resistance for LTOA samples using a semi-circular bending (SCB-IFIT) test (*10*).

To maintain consistency, samples from each laboratory were prepared for testing using the same procedure developed for reheating and aging. Following the same procedure eliminated the chance for bias to be introduced into the gathering of performance data.

Theoretical Maximum Specific Gravity Results

Table 2 shows the Gmm results of STOA and LTOA samples of the three ovens. One of the STOA replicates for Oven-A was damaged during the testing and the data point was removed from the analysis. The results indicate that the variability of the Gmm as per ASTM D 6857 appears to increase with increased aging because two of the three ovens exhibited a variance (d2s) greater than the allowable limit. This did not occur for the STOA samples; therefore the variability is as a result of the aging procedure, or the test procedure becomes more difficult to run, attesting the variability to operator error. It is worth mentioning that the LTOA protocol increased the Gmm values, which is logical and confirms the effect of more absorption during the long-term aging of the mixtures. When the absorption occurs, asphalt binder replaces air volume in the sample, leading to a higher Gmm value.

Hamburg Wheel-Tracking Testing Results

Compacted asphalt mixtures that undergo the HWT test can help predict failure susceptibility to permanent deformation (rutting) and moisture damage. In this study, samples were conditioned in a 46°C water bath for 30 minutes before running the test for 20,000 passes (52 passes/min), per standard WisDOT procedure. The rut depth data is recorded during the test by the linear variable differential transformers at the side of the steel wheel. As shown in Table 3, the average rut depth results were very similar among replicates between the different ovens. The minor variability among samples is to be expected, and a similar variability was noted in a previous study of Wisconsin mixtures (*11*). The coefficient of variance values for each study are within 1% of one another.

SCB-IFIT Testing Results

In the SCB-IFIT test, a semi-circular specimen with a 15-mm notch depth is subjected to a constant load in a three-point bending load configuration until fracture occurs. Strain

Table 3. HWT Results for the STOA Samples per Oven Type

	Rut depth at 20,000 Passes, mm			Creep slope, mm/pass		
Oven type	Sample	Average	COV%	Sample	Average	COV%
Oven-A	5.43	5.41	4.2	0.45	0.41	6.6
	5.40			0.37		
Oven-C	5.39	4.97		0.44	0.38	
	4.56			0.32		
Oven-B	5.62	5.22		0.37	0.36	
	4.81			0.34		

Table 4. Volumetric Properties Measured for the $LTOA_{18}$ Samples per Oven Type

Sample ID	Width, mm	Thickness, mm	Height, mm	Notch, mm	Gmb	Gmm	Air void, %
Oven-A-1	149	50	76	15.5	2.314	2.499	7.5
Oven-A-2	150	50	78	15.5	2.318		7.3
Oven-A-3	149	49	68	15.5	2.313		7.5
Oven-A-4	150	51	70	15.5	2.323		7.1
Oven-C-1	150	49	73	15.5	2.324	2.498	7.0
Oven-C-2	150	51	72	15.5	2.310		7.6
Oven-C-3	150	49	75	15.5	2.312		7.5
Oven-C-4	150	49	71	15.5	2.307		7.7
Oven-B-1	150	51	75	15.5	2.326	2.494	7.7
Oven-B-2	150	51	71	15.5	2.306		7.6
Oven-B-3	151	51	75	15.5	2.288		8.3
Oven-B-4	150	51	72	15.5	2.310		7.4

energy is computed at the notch of the specimen by recording the load and deformation. The computed strain energy is used to compare the cracking resistance of asphalt mixtures to meet the volumetric requirements of differing traffic levels tested at intermediate temperatures. In this study, four $LTOA_{18}$ replicate samples from each oven were tested under a loading rate of 50 mm/min after 2 hours of conditioning at 25°C. The volumetric properties measured for each replicate are included in Table 4. As shown, the sample air voids, height, and thickaness varied in a very narrow range. The raw data were analyzed using the SCB-IFIT software to determine the fracture energy (G_f) and the flexibility index (FI) for each sample. Table 5 shows the results of the SCB-IFIT tests conducted for samples from each oven. The raw values reported indicate significant replicate variability that have an effect on the FI value. The testing procedure followed (IFIT-405) (*12*) states that when four individual I-FIT specimens are tested, the FI value that is farthest from the average of the four may be discarded as an outlier to lower the variability of the average FI value that is reported. Regarding the results in Table 5, one sample from each oven was discarded, in this case sample 1 from Oven-A and Oven-C and sample 4 from Oven-B.

By performing analysis of variance on the FI results (α = 0.05), there is no significant statistical difference between the results of the different ovens. In addition, the FI results for all mixture SCB samples were found to be at or below 2.0. The low flexibility values can most likely be attributed to the aging condition the SCB-IFIT samples were subjected to. Figure 8 shows results from a recent study in Illinois (*13*) in which many projects reported values of the FI value at or below two; the samples in Figure 8 were subjected to 12 hours of aging as opposed to 18 hours.

Although a relatively high variability was noted for the samples tested in this study, the values obtained for FI are on the same order as those reported in other studies. The 18-hour aging duration used in the present study may be too harsh to obtain reasonable differentiation between mixtures, but it is nevertheless concluded that the oven type did not significantly affect the results of the SCB testing in this study.

Conclusions

In this study, the influence of HMA sample size, oven type, and variation in reheating/aging temperatures on the results of performance tests for a selected mixture was investigated. The samples were tested against rutting and cracking using the HWT and SCB-IFIT tests, respectively. As per the results presented in this study, the following conclusions can be made:

Table 5. SCB-IFIT Slope, Fracture Energy, and FI Results Measured for the $LTOA_{18}$ Samples per Oven Type

Sample ID	Slope	Avg.	COV%		Fracture energy, Joules/ m^2	Avg.	COV%		FI	Avg.	COV%	
Oven-A-1	−19.34	−10.1	7.6	10.8	1440.3	1229.4	26.9	17.5	0.7	1.2	33.8	24.3
Oven-A-2	−9.68				1607.6				1.7			
Oven-A-3	−10.99				998.6				0.9			
Oven-A-4	−9.65				1081.9				1.1			
Oven-C-1	−26.51	−11.2	8.9		1043.6	1131.7	10.0		0.4	1.0	14.8	
Oven-C-2	−12.36				1048.1				0.9			
Oven-C-3	−10.68				1259.8				1.2			
Oven-C-4	−10.59				1087.1				1.0			
Oven-B-1	−11.94	−10.5	19.4		1082.7	1055.3	3.7		0.9	1.0	14.1	
Oven-B-2	−9.06				1027.9				1.1			
Oven-B-3	Damaged				–				–			
Oven-B-4	−5.71				1156.4				2.0			

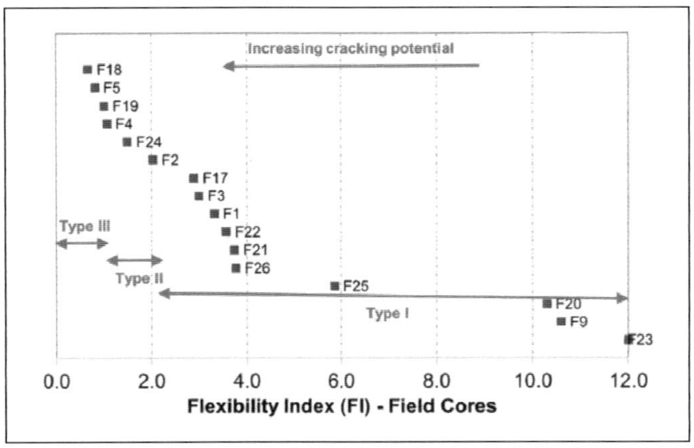

Figure 8. Flexibility index values for field cores from Al-Qadi et al. (*13*).

- Mixtures reach target aging temperatures quicker when there is no barrier between the mixture and the oven air, although the exposure of the mixture to moving air should be considered.
- The extent of oven loading can significantly affect the time it takes mixtures to reach designated aging temperature and can also affect the mixture temperature distribution within the oven.
- Briefly opening the oven to stir the sample during aging causes a reduction in mixture temperature, but in this study, it did not result in the mixture temperature falling below the specification guidelines used.
- The accuracy of the oven, as simulated by manually adjusting the aging temperature between the lower and upper limits of the specification used, did not show a significant effect on mixture volumetrics and performance for the samples and test procedures used in this study.
- Oven type did not affect the Gmm results significantly. However, the long-term aging increased the Gmm values and confirmed the effect of aging on mixtures' properties as the results became more variable and outside the limits stated in ASTM D 6857.
- No substantial influence of the oven type was found on the rut depth results obtained from the HWT test.
- As per the testing results in this study, resistance to fatigue cracking was significantly affected by the variability among the samples for each oven type. However, it appears that the 18 hours of long-term aging used in this study reduced the differences in cracking resistance between mixtures and may be too severe if differentiation between mixtures is desired.

The results of this study represent efforts to standardize mixture-handling and aging procedures within Wisconsin, but it is envisioned that such efforts are needed on different mixture types and on a national scale as well. Although it is unknown at this time whether the aging methods (loose versus compacted) and durations used in this study accurately simulate field conditions, it is anticipated that calibration of these procedures with field data will be necessary and is the focus of current research.

Acknowledgments

This study was funded by the Wisconsin Highway Research Program. Support of the project oversight committee is acknowledged.

Author Contributions

The authors confirm contribution to the paper as follows. Study conception and design: Hussain Bahia, Daniel Swiertz, Signe Reichelt; data collection: Daniel Swiertz, Zachary Lemke; analysis and interpretation of results: Zachary Lemke, Husam Sadek, Daniel

Swiertz, Hussain Bahia; draft manuscript preparation: Zachary Lemke, Husam Sadek. All authors reviewed the results and approved the final version of the manuscript.

References

1. Newcomb, D., A. Martin, F. Yin, E. Arambula, E. Park, A. Chowdhury, R. Brown, C. Rodezno, N. Tran, E. Coleri, D. Jones, J. T. Harvey, and J. Signore. *Short-term Laboratory Conditioning of Asphalt Mixtures*. NCHRP 815, National Cooperative Highway Research Program, Washington D.C., 2015.
2. Kim, Y., C. Hintz, F. Rad, S. Underwood, and M. Farrar. *Interim Report Literature Review*. NCHRP 09-54, National Cooperative Highway Research Program, Washington, D.C., 2013.
3. Zeinali, A., P. Blankenship, and K. Mahboub. Effect of Laboratory Mixing and Compaction Temperatures on Asphalt Mixture Volumetrics and Dynamic Modulus. *Transportation Research Record*: *Journal of the Transportation Research Board,* 2014. 2447: 101–108.
4. Hanz, A., E. Dukatz, and G. Reinke. Use of Performance Based Testing for High RAP Mix Design and Production Monitoring. *Association of Asphalt Paving Technologists, Road Materials and Pavement Design*, Vol. 18, 2016. pp 284–310.
5. TRB's National Cooperative Highway Research Program. *NCHRP Report 815: Short-term Laboratory Conditioning of Asphalt Mixtures*. Transportation Research Board of the National Academies, Washington D.C., 2015.
6. Elwardany, M., F. Rad, C. Castorena, and Y. Kim. Evaluation of Asphalt Mixture Laboratory Long-Term Aging Methods for Performance Testing and Prediction. *Association of Asphalt Paving Technologists, Road Materials and Pavement Design*, Vol. 18, 2016, pp 28–61.
7. ASTM D6857/D6857M-17 Standard Test Method for Maximum Specific Gravity and Density of Asphalt Mixtures Using Automatic Vacuum Sealing Method, ASTM International, West Conshohocken, PA, 2017, https://doi.org/10.1520/D6857_D6857M-17.
8. AASHTO T331. (2010). Standard Method of Test for Bulk Specific Gravity (Gmb) and Density of Compacted Hot Mix Asphalt (HMA) Using Automatic Vacuum Sealing Method. Washington D.C.: American Association of State Highway and Transportation Officials.
9. AASHTO T324. (2014). Hamburg Wheel-Track Testing of Compacted Hot Mix Asphalt (HMA). Washington D.C.: American Association of State Highway and Transportation Officials.
10. AASHTO TP124. (2016). Standard Method of Test for Determining the Fracture Potential of Asphalt Mixtures Using Semicircular Bend (SCB) Geometry at Intermediate Temperature. Washington D.C.: American Association of State Highway and Transportation Officials.
11. Bahia, H., P. Teymourpour, D. Swiertz, C. Ling, R. Varma, T. Mandal, P. Chaturabong, E. Lyngdal, and A. Hanz. *Analysis and Feasibility of Asphalt Pavement Performance-Based Specifications for Wisdot*. Project: 0092-15-04, Wisconsin Highway Research Program, University of Wisconsin-Madison, Madison, WI, 2016.
12. Circular Letter 2015-19. *Illinois Flexibility Index Test – Pilot Projects*. Illinois Department of Transportation, Illinois, 2015.
13. Al-Qadi, I. L., H. Ozer, J. Lambros, A. El Khatib, P. Singhvi, T. Khan, and B. Doll. *Testing Protocols to Ensure Performance of High Asphalt Binder Replacement Mixes Using RAP and RAS*. Illinois Center for Transportation, Illinois Department of Transportation, Illinois, 2015.

The Standing Committee on Quality Assurance Management (AFH20) peer-reviewed this paper (18-02799).

The results and opinions presented are those of the authors and do not necessarily reflect those of the sponsoring agency.

Article

TRR
JOURNAL OF THE TRANSPORTATION RESEARCH BOARD

Transportation Research Record
2018, Vol. 2672(26) 134–143
© National Academy of Sciences:
Transportation Research Board 2018
Article reuse guidelines:
sagepub.com/journals-permissions
DOI: 10.1177/0361198118793275
journals.sagepub.com/home/trr
SAGE

Effects of Mineral Additives on Cold Recycled Foamed Asphalt Mixtures Cured in Laboratory and Field Conditions

Wangyu Ma[1], Randy C. West[1], Nam Tran[1], Brian K. Diefenderfer[2], and Chen Chen[1]

Abstract

Cold recycling with foamed asphalt has been considered a sustainable pavement rehabilitation technology. During the mix design process, the dry and wet indirect tensile strengths (ITS) and tensile strength ratio (TSR) are important parameters for evaluating performance of a cold mix. These performance parameters can be significantly influenced by the type of mineral additive used in the mixture and conditions in which test specimens are cured before testing. In this study, the objective was to determine the effects of mineral additives on the dry and wet ITS and TSR results of the cold recycled foamed asphalt mixtures in laboratory- and field-curing conditions. Mixtures with four different additives (cement, hydrated lime, baghouse fines, and Class C fly ash) at varied contents were compared to a control mixture without a mineral additive. Specimens were cured in the laboratory and in a field-simulated condition. The results from laboratory curing show that cement and hydrated lime improved both the dry and wet ITS results. Baghouse fines increased the dry ITS but not the wet ITS, while the fly ash only enhanced the wet ITS. Higher additive contents did not improve dry or wet ITS results, except for cement. The field-simulated curing results show all the additives improved the dry ITS over long periods of time but were not effective in the short term. The ITS results from long-term field-simulated curing and laboratory curing ranked these mixtures similarly. Confinement during curing did not have a significant effect on the dry ITS.

Cold recycling has been used as a sustainable pavement rehabilitation technology because it uses reclaimed asphalt pavement (RAP) as the primary component, saving virgin aggregate and energy. Cold recycled mixtures are produced at the ambient temperature, significantly reducing greenhouse gas emissions generated during construction. Foamed asphalt or emulsified asphalt may be used as the binding agent for cold recycled mixtures. Foaming asphalt expands its volume and reduces its viscosity to disperse easily among RAP particles and form internal bonding in the mixture (*1*). In addition to RAP, foamed asphalt or asphalt emulsion, and water, a cold recycled mixture also has a mineral filler component to enhance its strength and reduce its moisture susceptibility (*2–4*). The most common mineral additive in cold recycling is Portland cement, while hydrated lime has been suggested as an alternative additive (*2*, *3*). Fly ash and limestone fines have also been mentioned as potential additives (*5*).

Cement and hydrated lime are commonly added up to 1.5% of the mixture. A few studies have compared different additives and suggested the optimum additive content for cold recycled asphalt mixtures. Nosetti et al. (*6*) compared the effects of cement and hydrated lime on the indirect tensile strength (ITS) of cold recycled mixtures using foamed asphalt and emulsified asphalt as recycling agents. They found mixtures with 0.5% of additives had much higher wet ITS and tensile strength ratio (TSR) than those with no additive. But improvements in ITS and TSR were limited when additive content increased from 0.5% to 2.0%. Mixtures with hydrated lime had equivalent ITS and TSR to the mixtures with cement.

[1]National Center for Asphalt Technology, Auburn, AL
[2]Virginia Transportation Research Council, Charlottesville, VA

Corresponding Author:
Address correspondence to Wangyu Ma: wzm0006@auburn.edu

Thanaya et al. (*7*) pointed out that adding 1%–2% cement significantly increases early-age and long-term strengths of cold recycled emulsified asphalt mixtures.

Use of fly ash in cold recycled mixtures is limited but it is still a potential alternative because of its lower cost. Cross and Young (*8*) verified the improvement in the moisture damage resistance of cold recycled emulsified asphalt mixtures by using fly ash. Fly ash production across the U.S. in 2015 was about 40 million metric tons with only 50% being used (*9*). Available survey data shows the price of fly ash was $74 per metric ton in 2013, about 33% lower than the price of cement ($111), and 49% lower than the price of hydrated lime ($144) based on the 2016 data (*10–12*). Another alternative additive is asphalt plant baghouse fines. The mineralogical composition of baghouse fines will depend on the sources of aggregates used in the asphalt plant. Eller and Olson (*5*) suggested limestone fines as a mineral filler for cold recycled mixtures.

Most studies on cold recycling additives were based on test results obtained after laboratory curing. However, in-situ field-curing conditions may be much different. Maccarrone (*2*) investigated curing condition of the recycled mixture and found that specimens being cured at 60°C for 3 days had similar moduli to those being cured in the field for 12 months. Ruckel et al. (*13*) found that laboratory curing at 40°C for 3 days simulates field curing condition in the range of 30–200 days. More studies are needed to compare test results obtained from laboratory- and field-curing conditions.

This study was meant to evaluate the effects of different mineral additives (cement, hydrated lime, baghouse fines, and Class C fly ash) on the cold recycled foamed asphalt mixtures. ITS testing was performed after both laboratory- and field-curing conditions to compare the effects of different curing factors (curing time, curing moisture, top load, and confinement). The top load was to simulate the mass of the hot mix asphalt (HMA) surface layer placed on top of the cold recycled pavement layer. In this study, HMA cores with the same mass and height were placed on top of the cold recycled specimens to apply the top load. Wired mesh was tied around each specimen to provide confinement.

Objectives and Scope

The primary objective of this study was to determine the effects of different mineral additives on the ITS results of cold recycled foamed asphalt mixtures in the laboratory- and field-curing conditions. Effects of additive type, additive content, and curing moisture level were studied based on test results obtained after the laboratory curing conditions. A secondary objective was to evaluate the effects of curing time, top load and confinement based on the results after the simulated field curing.

Materials

All cold recycled mixtures used in this study had the same RAP, foamed asphalt, and foamed asphalt content. The only differences among these mixtures were the mineral additive type and additive content. RAP was obtained from the Interstate-81 site in Virginia which was used on three test sections in the 2012 National Center for Asphalt Technology (NCAT) test track (*14*). The RAP gradation was between the medium and coarse gradations recommended by Asphalt Recycling & Reclaiming Association with 19-mm nominal maximum aggregate size (*15*). The RAP binder content was 4.85%. Foamed asphalt was produced from a PG 58-34 binder using a Wirtgen laboratory foaming device. The expansion ratio was 19 times, and the half-life was 10 seconds.

Four additives selected for testing included Portland cement (Type I/II), hydrated lime (Type S), fly ash (Class C), and baghouse fines from a local asphalt plant. These additives are very fine with most particles passing 0.075-mm sieve. The properties of these additives were tested following the respective ASTM standards, including C150 for the cement, C618 for the Class C fly ash, and D242 for the hydrated lime and the baghouse fines. The main components of baghouse fines were silica (SiO_2) and dolomite ($CaMg[CO_3]_2$) determined based on the X-Ray diffraction analysis. These components were not active and had no potential reaction with other materials in the cold recycled mixtures. A control mixture with no additive was also tested for comparison.

Methodology

Specimens were divided into five curing schemes, including three laboratory-curing conditions and two simulated field-curing conditions, as shown in Table 1. All the specimens were compacted with the same compaction effort (30 gyrations in Superpave Gyratory Compactor using a 100-mm diameter mold). The ITS results of three replicates were determined for each mixture after being cured in each scheme. Cement and hydrated lime were added at three levels (by weight of the dry RAP). A maximum of 2.0% was used for these two additives because excessive content may be detrimental to the mixture's flexibility (*4*). Mixtures with the Class C fly ash and baghouse fines cost much less than cement and hydrated lime; thus, these two additives were added up to 5.0% to explore the effect of additive content. At the higher contents, the two additives may affect the mixture flexibility, which should be further evaluated in the future. Additives were added uniformly and mixed sufficiently with RAP in a laboratory twin-shaft pug mill.

Scheme 1 is the most commonly used laboratory-curing method for cold recycled foamed asphalt mixtures. It consisted of curing specimens in a forced air draft oven at 40°C for 3 days. After curing, the ITS

Table 1. Experiment Plan

		Curing schemes				
		Lab			Field	
Additive	Content (%)	1	2	3	4	5
Control (CTRL)	—	3 days at 40°C	3 days at 40°C + 30 days in dry room[1]	3 days at 40°C + 30 days in moist room[1,2]	3, 14, 30, 100, (120) days outdoor[3,4]	100 days outdoor with top load and confinement[3,5]
Cement (CEM)	1.0, 1.5, 2.0					
Hydrated lime (HL)	1.0, 1.5, 2.0					
Class C fly ash (FA)	1.0, 1.5, 2.0, 5.0					
Baghouse fine (BHF)	1.0, 1.5, 2.0, 5.0					

Note: Ambient temperature in Scheme 2 and 3 is 20°C–25°C. Specimens were dried before testing ITS after curing following Scheme 3. Schemes 4 and 5 are outdoor environment conditions in Auburn, Alabama from November 2016 to February 2017; it only applies for the mixtures with 1.0% additive. The specimens were covered on rainy days. Specimens cured for 120 days in Scheme 4 were tested for wet ITS only to evaluate the moisture susceptibility. Wired mesh was tied around each specimen by hand in a consistent manner to provide confinement.

results of specimens under dry and wet conditioning were obtained in this scheme. The dry ITS results were obtained from specimens conditioned in dry environment for one hour at 25°C. The wet ITS results were obtained from those immersed in water for 24 hours at 25°C. Effects of additive type and additive content were determined for this scheme. Effect of curing moisture was determined by comparing the dry ITS of specimens cured in Schemes 2 and 3. Each scheme consisted of an initial curing (40°C for 3 days) and an extended curing in either dry or moisture room. After initial curing, specimens were cured in a dry laboratory room without moisture under Scheme 2. For Scheme 3, specimens were cured from the same batch in a moisture room of 100% humidity where moisture was constantly provided. This initial curing dried specimens and helped developing initial strengths to avoid damage during handling. However, in this step, moisture in the specimens evaporates quickly limiting the reaction between water and additives. Cardone et al. (*16*) suggests that these incomplete reactions may not adequately represent the material's strength. In Scheme 4 and 5, specimens were placed outdoor on a pallet with sufficient sunlight and wind to simulate field curing. The top load and confinement were used to simulate an asphalt overlay on the recycled pavement. Because the side and bottom of specimens were not sealed and still exposed to sunlight and wind, Scheme 5 only partially simulated the effect of asphalt overlay in the actual pavement condition. Only 1.0% additive was added in the field-curing specimens in Schemes 4 and 5. Figure 1 shows the two schemes in the field-curing condition.

Results and Analysis

Laboratory Curing

Effects of Additive Type and Additive Content. Figure 2 shows the average dry and wet ITS results of specimens cured at 40°C for 3 days (Scheme 1). Whiskers represent one standard deviation above and below the average. Figure 2*a* shows the results of both dry and wet ITS increased as more cement was added. The analysis of variance (ANOVA) showed that both cement content and testing condition (dry vs. wet) were significant effects on ITS (the *p*-value equals 0.007 and 0.001, respectively). A significance level (α) of 0.05 was used (*17*). Figure 2*b* shows the dry ITS results were a little higher at 2.0% hydrated lime than at other contents. Wet ITS slightly decreased as more hydrated lime was added. ANOVA results indicated that the effect of hydrated lime content was not significant, but the testing condition was (the *p*-value equals 0.945 and 0.001, respectively). Figure 2*c* shows the dry ITS results increased with baghouse fines content from 1.0% to 2.0% but decreased at 5.0% content. The TSR results ranged from 0.50 to 0.69, below that of cement (0.81–0.93) and hydrated lime (0.72–0.88). ANOVA on the results of three baghouse fines contents (1.0%–2.0%) showed that baghouse fines content was not a significant factor, but the testing condition was (the *p*-value equals 0.079 and less than 0.001, respectively). Figure 2*d* illustrates that either dry ITS or wet ITS result with no more than 2.0% Class C fly ash was lower than that at 5.0% content. Dry ITS decreased at 1.5% and increased at 2.0%. Wet ITS changed a little in the range of 1.0%–2.0%. The TSR result was higher than that of baghouse fines. ANOVA on the results of three Class C fly ash contents (1.0%–2.0%) showed the effect of the content was not significant on the dry and wet ITS, but the effect of testing condition was (the *p*-value equals 0.237 and less than 0.001, respectively). Figure 2*e* shows the TSR value of the control mixture was only 0.60, indicating the recycled mixture with no additive was moisture susceptible. T-test result verified the significant difference between dry and wet ITS of the control mixture (*p*-value equals 0.005) (*17*).

Figure 1. Field-curing condition: (*a*) specimens without top load and confinement and (*b*) specimens with top load and confinement.

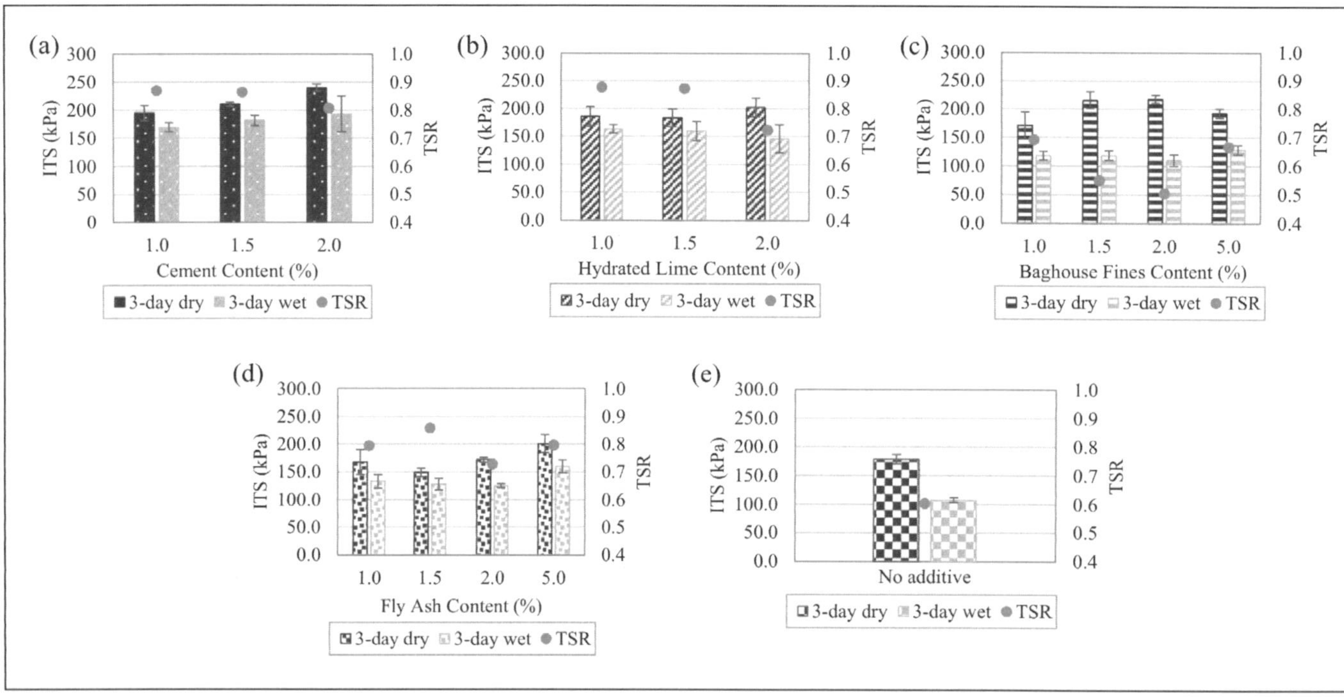

Figure 2. Comparison of ITS results of each mixture (Scheme 1): (*a*) cement; (*b*) hydrated lime; (*c*) baghouse fines; (*d*) Class C fly ash; (*e*) control.

To compare the effects of different additives further, the dry and wet ITS are shown separately in Figure 3. Figure 3*a* shows mixtures with cement (CEM), hydrated lime (HL), and baghouse fines (BHF) generally had higher dry ITS than those with Class C fly ash (FA) and the control mixture (CTRL). At 1.0% additive, the mixture with cement had the highest ITS, followed by mixtures with hydrated lime, baghouse fines, and the fly ash. At 1.5% additive, the mixture with baghouse fines had the highest ITS. At 2.0% additive, the mixture with cement became the highest again. In addition, increase of baghouse fines content to 5.0% did not improve dry ITS but increase of the fly ash to 5.0% did. The dry ITS of the control mixture was close to dry ITS of the mixtures with 1.0% baghouse fines or Class C fly ash. Figure 3*b* shows the differences of wet ITS results between five groups became more evident than those of dry ITS results. Mixtures with cement had the highest wet ITS at all three contents. The wet ITS of the control mixture was the lowest indicating the necessity of using additives to improve moisture susceptibility.

The mixtures with different additives were ranked regarding their dry and wet ITS in Table 2. The mixture with 2.0% cement had the highest dry and wet ITS. The

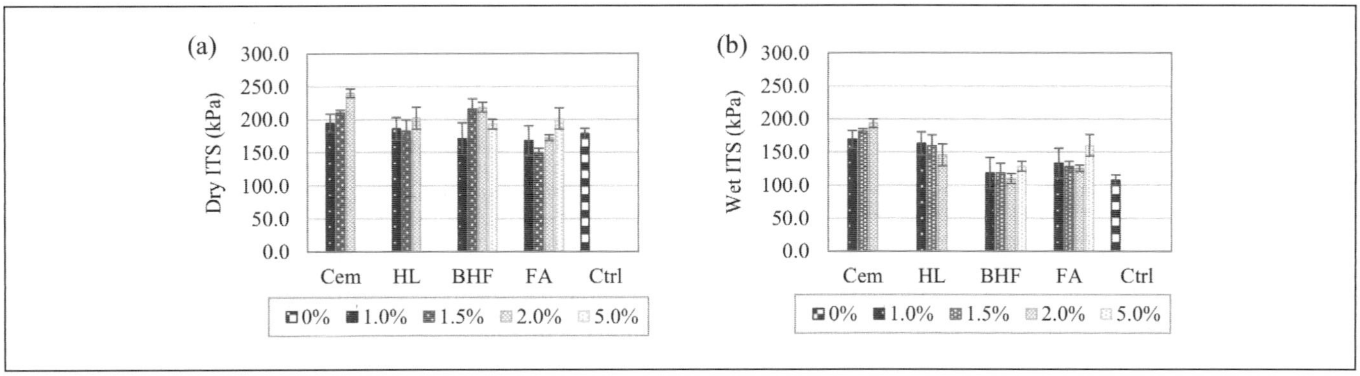

Figure 3. Comparing dry and wet ITS results of mixtures separately (Scheme 1): (*a*) dry ITS; (*b*) wet ITS.

Table 2. Numerical Ranking for Mixtures Based on the Dry and Wet ITS Results

	Dry ITS			Wet ITS		
Ranking	Additive type	Additive content (%)	Average (kPa)	Additive type	Additive content (%)	Average (kPa)
1	CEM	2.0	239.7	CEM	2.0	193.2
2	BHF	2.0	218.2	CEM	1.5	181.6
3	BHF	1.5	215.8	CEM	1.0	169.1
4	CEM	1.5	210.3	HL	1.0	163.2
5	HL	2.0	201.8	FA	5.0	160.1
6	FA	5.0	201.0	HL	1.5	159.7
7	CEM	1.0	194.7	HL	2.0	145.8
8	BHF	5.0	192.9	FA	1.0	133.4
9	HL	1.0	186.0	BHF	5.0	128.6
10	HL	1.5	183.0	FA	1.5	128.2
11	CTRL	0	178.6	FA	2.0	125.4
12	FA	2.0	172.3	BHF	1.5	118.6
13	BHF	1.0	171.0	BHF	1.0	118.4
14	FA	1.0	168.0	BHF	2.0	110.1
15	FA	1.5	149.5	CTRL	0	107.7

top six mixtures in dry ITS ranking include two mixtures with cement (2.0% and 1.5%), two with baghouse fines (2.0% and 1.5%), one with hydrated lime (2.0%) and one with Class C fly ash (5.0%). For the ranking by wet ITS, the top three are only mixtures with cement (1.0%, 1.5%, and 2.0%), followed by two with hydrated lime (1.0%) and one with Class C fly ash (5.0%). Mixtures with baghouse fines were ranked much lower based on wet ITS. Generally, mixtures with cement performed well in both dry and wet conditions. Mixtures with baghouse fines may have comparable dry ITS but become susceptible to moisture damage. Table 2 also shows the benefit of including mineral additives in the recycled mixture. The control mixture (no mineral additive) had the lowest wet ITS and a dry ITS among the lowest of all the mixtures studied.

Effect of Curing Moisture. Figure 4 compares the ITS results of specimens cured in dry and moist rooms, as well as the TSR calculated as the ratio of the dry ITS results from the two curing conditions (hereafter referred to as TSR'). After moist-room curing, specimens with cement have higher ITS than other mixtures. The values of TSR' for the mixture with baghouse fines are lower than those of other mixtures suggesting the potential problem of using baghouse fines when mixtures are exposed to excessive moisture.

Figure 4*a* shows the dry ITS increased as more cement was added. ANOVA results showed that both cement content and curing moisture have significant effects on dry ITS (both effects had the same p-value equals to 0.004). It means the specimens with cement cured in moisture room had statistically higher ITS than those cured in dry room. This result indicated that part of the unhydrated cement was left from initial curing hydrated and caused the dry ITS increase during moisture room curing. Figure 4*b* shows that the dry ITS from specimens cured in a moisture room is greater than dry ITS for those specimens cured in dry room only at 1.0%

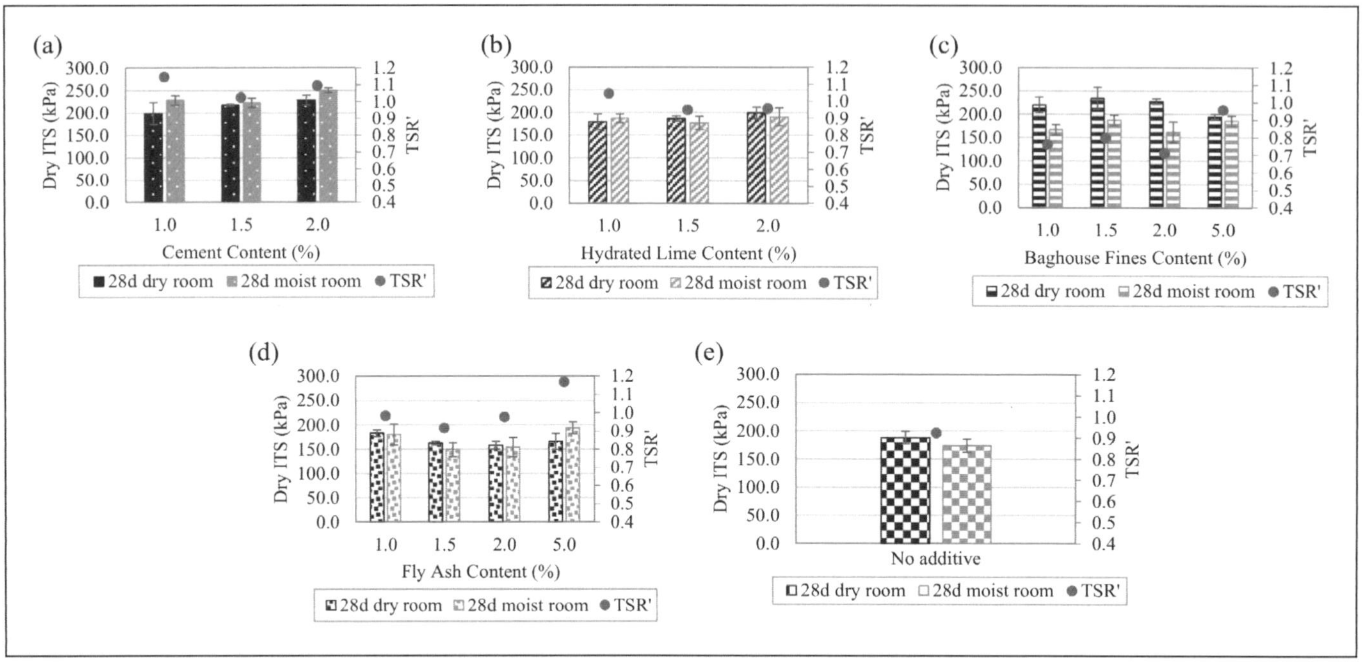

Figure 4. Comparison of the dry ITS results for mixtures cured in dry and moisture rooms (Schemes 2 and 3): (*a*) cement; (*b*) hydrated lime; (*c*) baghouse fines; (*d*) Class C fly ash; (e) control.

hydrated lime content. ANOVA showed the effects of both hydrated lime content and curing moisture were not significant (the *p*-value equals 0.235 and 0.599, respectively). Comparison of the dry ITS of mixtures with baghouse fines is shown in Figure 4*c*. Moisture room curing reduced the dry ITS at all four additive contents. The dry ITS reduction at 5.0% baghouse fines was smaller than the reductions at other contents. T-test results showed the effect of moisture curing on dry ITS was significant (the *p*-value was less than 0.001). In Figure 4*d*, the t-test results showed effect of curing moisture on the dry ITS of mixes with Class C fly ash was not significant (the *p*-value equals 0.908). Figure 4*e* shows specimens cured in moisture room have lower dry ITS. However, t-test showed the ITS results tested in two curing schemes are statistically the same (the *p*-value equals 0.198).

Field Curing

Effect of Curing Time. The effect of field-curing time was evaluated by measuring the dry ITS of specimens cured for different periods of time (3, 14, 30, and 100 days) and wet ITS after 120 days. For specimens cured longer than 14 days, the exact testing date was controlled within ± 3 days of the plan. For those cured for 100 days, testing date was controlled within ± 5 days of the plan. Specimens with the same additive were produced from one batch. Figure 5 shows the development of dry ITS for the five mixtures after curing. The coefficients of variance of the dry ITS results at different testing dates were between 6.3% to 12.7%. The dashed horizontal line in each plot indicates dry ITS tested after curing Scheme 1 (3 days at 40°C).

Figure 5*a* shows the dry ITS of the mixture with cement. Approximately 92% of the 100-day dry ITS occurred during the first 30 days. The dry ITS after curing at 40°C for 3 days was between the 30-day and 100-day field-curing strength. Figure 5*b* shows that the dry ITS of mixture with hydrated lime increases sharply from 3 to 14 days, and then decreased slightly after 30 days. This unexpected lower dry ITS may have been due to material variability such as nonuniform distribution of the additive, RAP gradation, or asphalt. A greater increase in the dry ITS occurred between 30 days and 100 days than did in the first 2 weeks. The laboratory-cured dry ITS was between the 30-day and the 100-day field-cured strengths. Figure 5*c* shows the dry ITS of mixtures with baghouse fines increased rapidly within first two weeks, then followed by a lesser increase until 30 days. The laboratory-cured specimens had dry ITS value close to the 14-day field-cured specimens. The dry ITS of the mixtures with Class C fly ash is shown in Figure 5*d*. The dry ITS increased at a higher rate in the first month and then was consistent after the first month. Laboratory-cured specimens had dry ITS value slightly higher than 14-day field-cured specimens. Figure 5*e* shows that the dry ITS of the control mixture increased in the first 2 weeks, stayed constant after until 30 days, and decreased a little afterwards. However, comparison

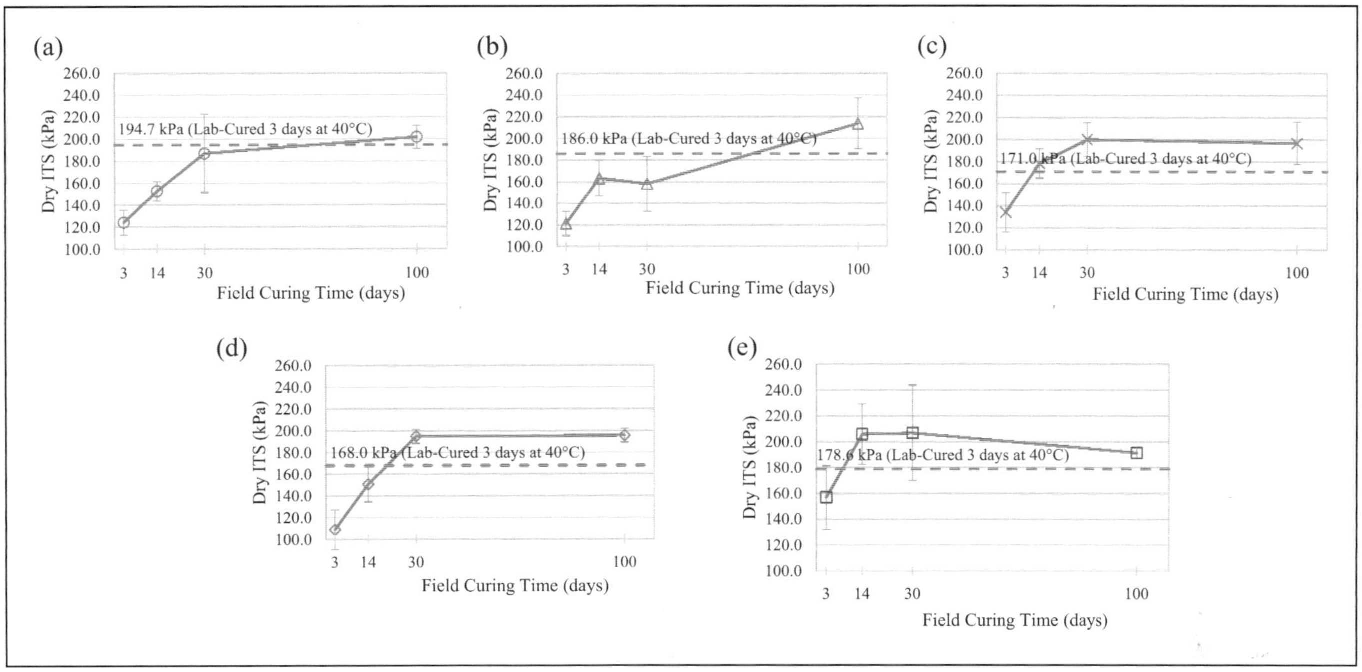

Figure 5. Dry ITS of mixtures after different periods of field curing (Scheme 4): (a) cement; (b) hydrated lime; (c) baghouse fines; (d) Class C fly ash; (e) control.

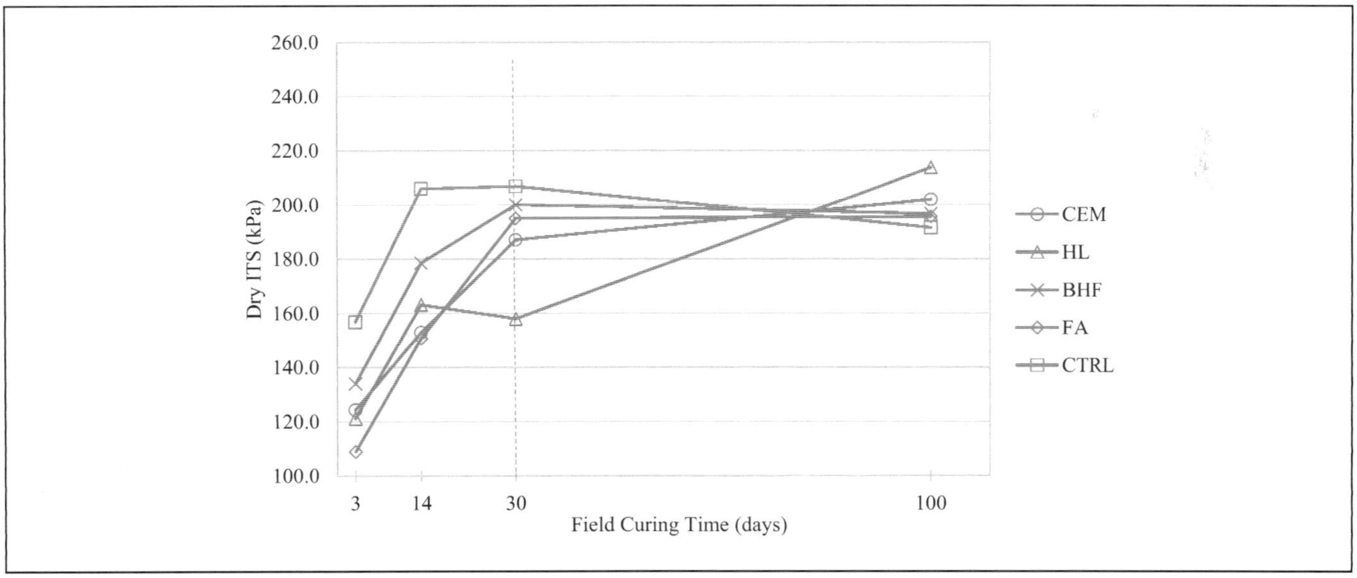

Figure 6. Comparison of dry ITS for mixtures with different additives after field curing (Scheme 4).

between 30- and 100-day dry ITS results by t-test showed these two did not have a statistically significant difference (the p-value equals 0.550). The unexpected decrease from 30 days to 100 days may have been due to material variability. For the control mixture, laboratory-cured dry ITS was equivalent to the field-cured strength between 3 days and 14 days.

Furthermore, Figure 6 shows dry ITS of these five mixtures after field curing for convenience of comparing. Each line represents development of dry ITS for a mixture with one additive. All the mixtures have sharp increases of dry ITS from 3 to 14 days in similar rates. Then, most mixtures (except the mixture with hydrated lime) showed slower increase from 14 to 30 days.

Table 3. Numerical Ranking and Tukey's Test Grouping Based on Dry ITS Results after Field Curing

	Numerical ranking				Tukey's test grouping			
Additive in the mixture	3-Day	14-Day	30-Day	100-Day	3-Day	14-Day	30-Day	100-Day
Cement	3rd	4th	4th	2nd	A B	B	A	A
Hydrated lime	4th	3rd	5th	1st	A B	A B	A	A
Baghouse fines	2nd	2nd	2nd	3rd	A B	A B	A	A
Class C fly ash	5th	5th	3rd	4th	B	B	A	A
Control	1st	1st	1st	5th	A	A	A	A

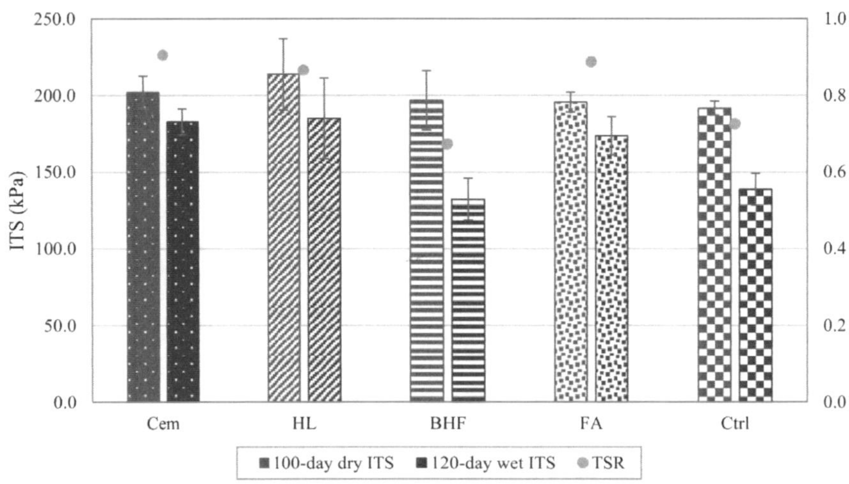

Figure 7. Comparison of dry and wet ITS after long-term field curing (Scheme 4).

Mixtures with cement and Class C fly ash had increased at the same rate until 30 days; mixture with baghouse fines had lower rate of increase; increase of control mixture almost stopped. The differences of 30-day dry ITS among the four mixtures with additives became smaller. After 30 days, these mixtures changed a little except for hydrated lime. After 100 days, the five mixtures had similar dry ITS in the range of 192–202 kPa. All the mixtures with additives had higher dry ITS than the control mixture.

The five mixtures were ranked numerically and compared pairwise based on dry ITS results after every field curing period. The ranking and comparison results from Tukey's test are summarized in Table 3 (*17*). Considerable change in numerical ranking takes place in the mixture with hydrated lime (from 5th to 1st), control mixture (from 1st to 5th), and mixture with cement (from 4th to 2nd). The 100-day ranking of these mixtures is close to the numerical ranking of laboratory-cured mixtures (1.0% additive) shown in Table 2. Numerically, mixtures with cement and hydrated lime have higher dry ITS than those with baghouse fines and Class C fly ash; while the control mixture was ranked differently between 100-day field curing and laboratory curing. The Spearman's rank correlation coefficient (ρ) equals 0.60 indicating good positive correlation (*18*). The two similar rankings suggest that the dry ITS test results for mixtures cured for 3 days at 40°C can be used to rank the relative strengths of long-term cured mixtures (longer than 100 days). Tukey's test results show that the dry ITS of the five mixtures were grouped into only two categories at early ages (3 days and 14 days) after which there was no statistical difference in the dry ITS value for field-cured mixtures.

However, examining only dry ITS results would miss the potential for damage from exposure to moisture. To compare these mixtures' moisture susceptibilities, additional specimens (three replicates for each mixture) were cured for 120 days before conducting wet ITS tests. Figure 7 shows the 120-day wet ITS of the mixtures compared to the dry ITS results tested after 100 days of field curing. The ratio of 120-days wet ITS to the 100-day dry ITS (hereafter referred to as TSR) is also shown as dots. The cement, hydrated lime, and Class C fly ash mixtures showed slight reductions in wet ITS values (TSR were all greater than 0.87) while more significant reductions were evident in the mixture with baghouse fines (TSR = 0.67)

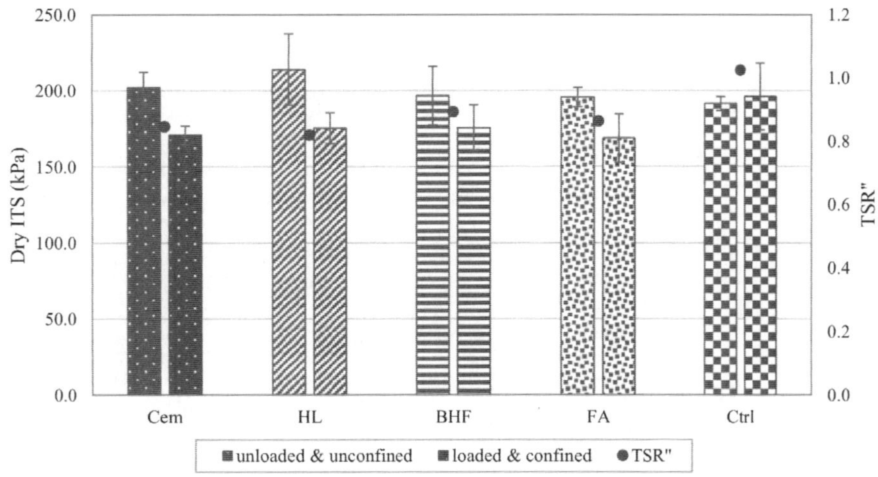

Figure 8. Comparison of dry ITS for specimens with and without top load and confinement after field curing 100 days (Schemes 4 and 5).

and the control mixture (TSR = 0.72). The three mixtures using cementing additives were those that had the highest TSR. Rankings of these mixtures (with 1% additive) by wet ITS were also similar to the rankings in Table 2 (ρ = 0.80). This suggests that ITS results for mixtures cured for 3 days at 40°C may be used to rank the relative strengths of long-term cured mixtures (100 days). The similar rankings of dry or wet ITS results between the lab-cured (3 days at 40°C) and the long-term field-cured specimens indicated possible strong correlation. To validate, the correlation of ITS results between two curing conditions was determined and the R-squared value of linear regression was 0.88, indicating strong positive correlation.

Effect of Top Load and Confinement. Figure 8 shows the comparison of dry ITS for specimens with and without top load and confinement. All the mixtures with additives showed a decrease in dry ITS because of the top load and confinement (the ratio of these two dry ITS results were referred to as TSR"). The specimens in confinement were still exposed to the sunlight and wind except for the top surface. They were expected to have higher ITS because of densification caused by the applied top load and confinement. However, it turned out that this effect was not significant. Therefore, the covered top still reduced the curing rate of additive and had more significant impact on the strength development than the densification effect. However, t-tests showed that the effect of top load and confinement was only significant for the mixture with cement (the p-value equals 0.020).

Findings and Conclusion

The study examined the effects of different mineral additives on the ITS of cold recycled foamed asphalt mixtures subjected to laboratory- and field-curing conditions. Based on the test results presented above, the following findings and conclusions can be made:

- After the commonly used laboratory-curing condition (3 days at 40°C), mixtures with cement and baghouse fines had higher dry ITS than mixtures with hydrated lime, Class C fly ash, or no additive. Also, mixtures with cement and hydrated lime had higher wet ITS. Cement was the best overall mineral additive considering improvements to both dry and wet ITS.
- Higher additive contents do not necessarily improve the dry and/or wet ITS results of mixtures. Only ITS of the mixture with cement increased as more cement was added. However, the effect of higher cement contents (more than 2%) on the mixture flexibility was not measured in this study, and it could affect the recommended dosage.
- Curing moisture had a significant effect on a mixture's dry ITS when cement or baghouse fines was used. Sufficient moisture helped increase hydration of cement, but it significantly reduced the dry ITS of mixture with baghouse fines.
- Dry ITS of all mixtures with and without mineral additives increased with field curing time. After 30 days of field curing, changes in dry ITS were minor for mixtures with cement, baghouse fines, Class C fly ash, and the control mixture. Mixtures with hydrated lime required more than 100 days to develop equivalent strengths as other additives. Comparing laboratory- and field-curing results, test results of mixtures cured for 3 days at 40°C provided consistent ranking as long-term curing (at least 100 days).

- Top load and confinement had a significant effect on the mixture with cement. For mixtures with the other additives, this effect on dry ITS was negligible.
- Due to their higher wet-ITS values, cement, hydrated lime, and Class C fly ash are better additives for foamed asphalt mixtures than baghouse fines or no additive where minimum wet ITS criteria and/or TSR criteria are used.

Author Contributions

The authors confirm contribution to the paper as follows: study conception and design: Wangyu Ma, Randy West; data collection: Wangyu Ma, Chen Chen; analysis and interpretation of results: Wangyu Ma, Randy West; draft manuscript preparation: Wangyu Ma, Randy West, Nam Tran, Brian Diefenderfer. All authors reviewed the results and approved the final version of the manuscript.

References

1. Chan, P., S. Tighe, and S. Chan. Exploring Sustainable Pavement Rehabilitation: Cold In-Place Recycling with Expanded Asphalt Mix. Presented at 89th Annual Meeting of the Transportation Research Board, Washington, D.C., 2010.
2. Maccarrone, S. Cold Asphalt Systems as an Alternative to Hot Mix. *Asphalt Review*, Vol. 14, No. 1, 1995, pp. 19–24.
3. *Technical Guideline: Bitumen Stabilised Materials*, 2nd ed. Asphalt Academy, Pretoria, South Africa, 2009.
4. Wirtgen GmbH. *Wirtgen Cold Recycling Technology*. Windhagen, Germany, 2012.
5. Eller, A. J., and R. Olson. *Recycled Pavements Using Foamed Asphalt in Minnesota*. Publication MN/RC 2009-09. *Minnesota Department of Transportation*, Saint Paul, MN, 2009.
6. Nosetti, A., F. Pérez-Jiménez, A. Martínez, and R. Miró. Effect of Hydrated Lime and Cement on Moisture Damage of Recycled Mixtures with Foamed Bitumen and Emulsion. Presented at 95th Annual Meeting of the Transportation Research Board, Washington, D.C., 2016.
7. Thanaya, I. N. A., S. E. Zoorob, and J. P. Forth. A Laboratory Study on Cold-mix, Cold-lay Emulsion Mixtures. *Proceedings of the Institution of Civil Engineers: Transport*, Vol. 162, No. TR1, 2009, pp. 47–55.
8. Cross, S. A., and D. A. Young. Evaluation of Type C Fly Ash in Cold In-Place Recycling. *Transportation Research Record: Journal of the Transportation Research Board*, 1997. 1583: 82–90.
9. American Coal Ash Association. *Fly Ash Production and Use with Percent*. https://www.acaa-usa.org/Portals/9/Files/PDFs/2015-Survey_Results_Charts.pdf. Accessed June 1, 2017.
10. Heavey, C., A. Hernandez, S. Hughes, and L. Willman. Comparative Life Cycle Assessment of Traditional Concrete and Concrete Made with Fly Ash. *Ash at Work*. No. 1, 2015, pp. 27–35. https://www.acaa-usa.org/Portals/9/Files/PDFs/ASH01-2015.pdf. Accessed June 5, 2017.
11. Statista. *Cement Prices in the U.S. from 2007 to 2016*. https://www.statista.com/statistics/219339/us-prices-of-cement/. Accessed May 10, 2017.
12. US Geological Survey. *Mineral Commodity Summary: Lime*. January 2017. https://minerals.usgs.gov/minerals/pubs/commodity/lime/mcs-2017-lime.pdf. Accessed March 8, 2017.
13. Ruckel, P. J., S. M. Acott, and R. H. Bowering. Foamed-Asphalt Paving Mixtures: Preparation of Design Mixes and Treatment of Test Specimens. *Transportation Research Record: Journal of the Transportation Research Board*, 1983. 911: 88–95.
14. Diefenderfer, B. K., M. Diaz-Sanchez, D. H. Timm, and B. F. Bowers. *Structural Study of Cold Central Plant Recycling Sections at the National Center for Asphalt Technology Test Track*. Report VTRC 17-R9. Virginia Transportation Research Council, 2016.
15. *Recommended Mix Design Guidelines for Cold Recycling Using Foamed (Expanded) Asphalt Recycling Agent (Draft)*. Asphalt Recycling & Reclaiming Association, Annapolis, MD, 2016.
16. Cardone, F., A. Grilli, M. Bocci, and A. Graziani. Curing and Temperature Sensitivity of Cement–Bitumen Treated Materials. *International Journal of Pavement Engineering*, Vol. 16, No. 10, 2015, pp. 868–880.
17. Montgomery, D. C. *Design and Analysis of Experiments*. John Wiley & Sons, New York, NY, 2017.
18. Royal Geographical Society. *Spearman's Rank Correlation Coefficient - Excel Guide*. http://www.rgs.org/NR/rdonlyres/4844E3AB-B36D-4B14-8A20-3A3C28FAC087/0/OASpearmansRankExcelGuidePDF.pdf. Accessed July 10, 2017.

The Standing Committee on Asphalt Pavement Construction and Rehabilitation (AFH60) peer-reviewed this paper (18-05819).

Article

TRR
JOURNAL OF THE TRANSPORTATION RESEARCH BOARD

Transportation Research Record
2018, Vol. 2672(26) 144–154

DOI: 10.1177/0361198118794068
journals.sagepub.com/home/trr
SAGE

Enhanced Model for Continuous Dielectric-Based Asphalt Compaction Evaluation

Kyle Hoegh[1], Shongtao Dai[1], Trevor Steiner[1], and Lev Khazanovich[2]

Abstract

The compaction of asphalt concrete significantly affects long-term pavement performance. Although coring provides a relatively accurate way of assessing in-place density at specific locations, the coverage of the assessment is limited, especially at longitudinal joint locations. This can be particularly problematic because it is difficult to identify problematic locations that are likely to fail prematurely using current compaction assessment methods. Ground penetrating radar (GPR) provides an attractive nondestructive testing alternative for evaluation of compaction quality, especially with recent significant improvements in the GPR technology for this specific application. However, assessment of the air void content of the asphalt mix from the GPR-measured dielectric constant of the surface requires conversion of dielectric variation to air void content variation, which is the subject of this paper. An alternative to the commonly used model is proposed, leading to more justifiable predictions for low values of dielectric constants. The proposed model was used to interpret data from a 7-mi long asphalt overlay construction project. The results of the interpretation as compared with the results obtained with the conventional model show an improvement on the stability of the prediction at low air void contents, especially when core calibration data are limited and uncertainty is considered. These results are promising in the direction of reducing field cores necessary to have a stable model providing continuous compaction assessment of new asphalt pavement construction.

Past research has shown that the performance of asphalt concrete is highly dependent on the air void content of the compacted asphalt mixture. The air void content has been shown to correlate with key asphalt characteristics such as stiffness (*1*), strength (*2*) and dynamic modulus (*3*). Kassem et al. found that increased air void content correlated with the increased occurrence of various pavement distresses including excessive aging and moisture damage that negatively affected long-term performance (*4*). The impacts on long-term performance were quantified in a study performed by Linden et al., which estimated that each 1% increase in air voids over 7% leads to an approximately 10% reduction in pavement life (*5*).

Typically, asphalt compaction is assessed using coring, which is destructive, expensive, time consuming, and limited in coverage. Though these measurements are useful for postconstruction analysis and are often used as primary components of quality assurance measurements, they cannot provide a real-time feedback during the paving operation. The issues associated with traditional measures of compaction create a need for nondestructive methods that can collect data continuously, cheaply, and quickly.

Ground penetrating radar (GPR) provides a nondestructive testing alternative that allows for walk-behind or vehicle-mounted measurements (*6*, *7*). GPR uses electromagnetic waves to explore subsurface characteristics. In transportation infrastructure survey, GPR has been commonly applied to detect free water (*8*), to estimate the dielectric property of pavement materials (*9*), to estimate the layer thicknesses (*10*), and to study the asphalt concrete (AC) layer density (*11–14*). ASTM standard ASTM D6432-11 provides a procedure of applying GPR for subsurface investigation.

Determination of dielectric properties of the asphalt layer with GPR has been traditionally done through measurement of either round trip travel time to reflection at the depth of the AC layer or surface reflection. The travel time approach covers a greater depth, but relies on a known thickness. The asphalt thickness is often

[1]MnDOT Materials and Road Research, Maplewood, MN
[2]University of Pittsburgh, Pittsburgh, PA

Corresponding Author:
Address correspondence to Kyle Hoegh: kyle.hoegh@state.mn.us

unknown and spatially variable. Moreover, if the asphalt layer is placed in several lifts or as an overlay over an existing asphalt pavement, it may be difficult to separate the travel time in the individual lifts from the overall travel time of the electromagnetic signal in the asphalt layer.

The AC surface reflection method uses the ratio of the amplitude of the GPR signal reflection from air to the asphalt surface, A_0, to the incident amplitude (represented by the reflection from the metal plate), A_i, to determine the bulk dielectric constant of the asphalt, e_r. The dielectric constant of the surface is determined according to Saarenketo and Scullion (*15*) using:

$$e_r = \left(\frac{1 + \left(\frac{A_0}{A_i}\right)}{1 - \left(\frac{A_0}{A_i}\right)} \right)^2 \tag{1}$$

The advantage of this approach is that if the upper lift is sufficiently thick (thicker than 30 mm) then the measured AC surface reflection depends only on the properties of the upper layer.

For newly placed asphalt lift, the dielectric constant values determined from Equation 1 can be empirically related to the relative ratio of pore volume to the total volume of each specific asphalt mix because air has a lower dielectric constant than the surrounding asphalt material and the aggregate type and volumetric proportion are typically uniform (*16*, *17*). Because dielectric properties of the asphalt mix depend on the dielectric properties of other components of the mix which vary from project to project a universal dielectric constant to air void content conversion is not feasible. To account for these changes, cores need to be taken for each new mix at locations where conversion from dielectric constant to air void content is conducted. The correlation between the air voids and dielectric constants plays a key role in the accuracy of the air voids predictions made using the model.

There have been several field trials of GPR for nondestructive testing (NDT) determination of air voids. The first large scale trial was performed in Finland in 1996–1997 (*18*). Recently, several state departments of transportation in the USA have held trials of the technology (*19–22*). The most notable recent application of GPR for compaction surveying was conducted by Sebesta, Scullion and Saarenketo (*16*) as part of the SHRP2 RO6C activities. Though several implementations have consistently demonstrated the ability of GPR technology to provide real-time information on relative compaction, use of GPR to estimate actual air void content has limitations and is still a challenging problem in spite of the significant improvements in the GPR technology for this specific application under the SHRP2 project (*16*).

Various impulse radar versions of GPR have shown that the dielectric properties determined from the asphalt surface reflection amplitude correspond with core-measured air void content (*18*, *22*, *23*). Additionally, a step frequency array-based method improves the coverage and productivity of the measurements, making it an attractive alternative to current state-of-the-practice procedures (*21*). Whereas these studies showed the potential of new technology for improved quality assurance in selected locations, the focus of this study is on how a stable compaction assessment can be achieved in full-scale implementation. In the case of the step frequency array system (*24*), these technologies can require intensive data processing from the frequency domain or can be cost prohibitive, whereas the single impulse array systems do not provide necessary coverage for widespread implementation.

The GPR equipment used in this study, the rolling density meter (RDM), is based on a system that evolved from recent research conducted under a National Academies of Science sponsored Strategic Highway Research Program (SHRP2) (*16*). It uses similar antennae, but also applied in a three-channel array to obtain some of the benefits in coverage explained in (*21*), in which multiple antenna pairs are used in each pass.

This paper deals with a problem of conversion of dielectric variation to air void content variation. Traditionally, a simple exponential model is used for dielectric–air void content correlation. The current practice is to develop this model for each construction project to minimize the error caused by variations in mix design and properties. To minimize extrapolation outside the calibration limits, it is recommended to collect cores representing full range survey dielectric constant values (*18*, *19*, *21*). However, this is not always practical because the full range of dielectrics on a project are not known until the project is complete. Moreover, areas with a high air void content often exhibit higher local variation causing higher uncertainty in the core air void measurements. Because only a limited number of cores with a high air void content are available this increases the uncertainty in the dielectric property–air void relationship for high air void content values. At the same time, an accurate conversion of the dielectric values to the air void content is especially important for the areas with low measured dielectric values, because they often determine whether the pavement will fail prematurely.

The paper presents an approach for development of a modified model to convert dielectric variation to air void content variation. The paper considers electromagnetic mix modeling theory and is based on the conventionally accepted empirical method. The proposed model includes physics-based constraints to ensure more reliable determination of the areas with high air void content.

Table 1. Mix Model Inputs (*23*)

Model		Aggregate (ε_s)	Binder (ε_b)	Maximum bulk SG (G_{mm})	SG of binder (G_b)	Binder content (P_b)	Bulk SG of aggregate (G_{sb})
Rayleigh	Assumed	6	3	2.521	1.015	5%	2.705
	Regression	7.44	2.00	2.276	1.015	4.1%	2.610
Bottcher	Assumed	6	3	2.521	1.015	5%	2.705
	Regression	8.33	7	3.026	1.015	6.0%	2.610
Crim	Assumed	6	3	2.521	1.015	5%	2.705
	Regression	6.38	3.21	2.602	1.015	5.1%	2.610

Note: SG = specific gravity.

Overview of Electromagnetic Mix Modeling Approach

Several approaches have been developed for relating the dielectric properties of an asphalt mix to its air void content. The most commonly used approach is the use of an empirical correlation between these two mix characteristics. Many studies successfully used the exponential model (conventional model) (*15, 16, 21, 22*):

$$AV = A^*\exp(-B^*\varepsilon) \tag{2}$$

where AV is the air void content, ε is the input hot mix asphalt dielectric constant, and A and B are calibration constants.

Although this purely empirical relationship is effective in converting dielectric constant to air void content when the necessary dielectric precision is achieved and sufficient core calibrations are available, a more rigorous approach should derive the air void–dielectric constant relationship for the asphalt mix from the dielectric properties and volume fractions of the asphalt mix components. Although this ideal is not fully achievable using the currently available mix models, the mix modeling approach should be considered to ensure a more stable physics-based empirical model.

Electromagnetic mix modeling has been developed for dielectric characterization of rock properties such as estimation of rock porosity (*25*), water and clay influences (*26*), and other geophysical applications (*27*). This work was further extended for application to civil engineering materials including evaluation of hot mix asphalt using GPR (*23, 28–30*). The mix model method is based on the modeling of how an electromagnetic wave interacts with composite materials using assumptions for how the wave interacts with the different AC components. The mix models evaluated in this study differ only in form, as the coefficients used (dielectric and volumetric values) are the same in each model. A more detailed description of each model can be found elsewhere (*23, 28–30*). Equations 3 through 5 show commonly accepted mix models for asphalt concrete.

Rayleigh Model:

$$AV = 1 - \frac{\frac{\varepsilon_{HMA}-\varepsilon_b}{\varepsilon_{HMA}+2^*\varepsilon_b} - \frac{1-\varepsilon_b}{1+2^*\varepsilon_b}}{G_{mm}^*\left(\left(\frac{\varepsilon_s-\varepsilon_b}{\varepsilon_s+2^*\varepsilon_b}\right)^*\left(\frac{1-P_b}{G_{sb}}\right) - \frac{(1-\varepsilon_s)}{1+2^*\varepsilon_b}{}^*\left(\frac{1}{G_{mm}}\right)\right)} \tag{3}$$

Bottcher Model:

$$AV = 1 - \frac{\frac{\varepsilon_{HMA}-\varepsilon_b}{3^*\varepsilon_{HMA}} - \frac{1-\varepsilon_b}{1+2^*\varepsilon_{HMA}}}{G_{mm}^*\left(\left(\frac{\varepsilon_s-\varepsilon_b}{\varepsilon_s+2^*\varepsilon_{HMA}}\right)^*\left(\frac{1-P_b}{G_{sb}}\right) - \frac{(1-\varepsilon_b)}{1+2^*\varepsilon_{HMA}}{}^*\left(\frac{1}{G_{mm}}\right)\right)} \tag{4}$$

CRIM for $\alpha = 2$:

$$AV = 1 - \frac{\sqrt{\varepsilon_{HMA}} - 1}{G_{mm}^*\left(\frac{P_b}{G_b}\sqrt{\varepsilon_b} + \frac{(1-P_b)}{G_{sb}}\sqrt{\varepsilon_s} - \frac{1}{G_{mm}}\right)} \tag{5}$$

where AV is the air void content, ε_{HMA} is the input hot mix asphalt dielectric constant, and the remaining variables are given in Table 1.

The mixing models are advantageous as compared with a purely empirical approach in that they are rational. These models—combined with laboratory testing and more accurate assumptions—may eventually reduce or even eliminate the need for calibration cores. These models guarantee reasonable trends—sensitivity to the properties of the individual components, that is—so it may make them less susceptible to errors. For example, each model can be verified for the extreme conditions, such as predicting 100% air void content (0 bulk specific gravity) when a dielectric constant of 1 is used. By comparison, the conventional purely empirical model does not hold true outside of the measured dielectric ranges.

Al-Qadi et al. conducted a sensitivity study and evaluated the CRIM, Rayleigh, and Bottcher models for characterizing asphalt compaction (*23*). They employed a method of calibrating the models using laboratory mixed and compacted specimens (*23*). Table 1 gives the assumed values used in the sensitivity study as well as the values obtained from a least squares regression matching the laboratory-measured air voids.

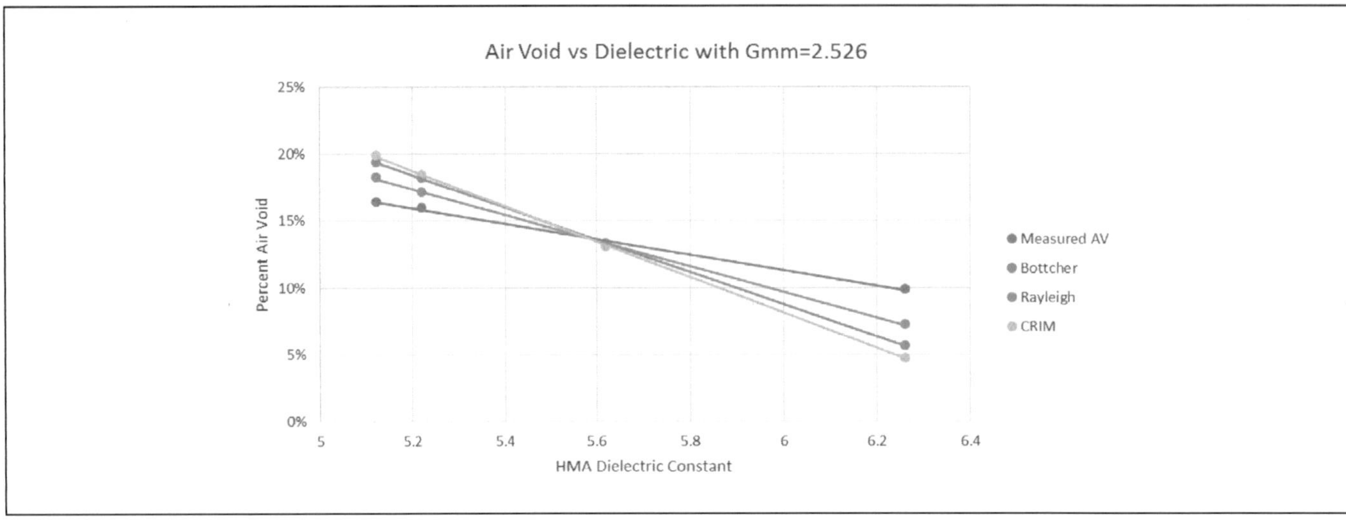

Figure 1. Measured (blue) versus predicted air void content using the CRIM, Rayleigh, and Bottcher models (*23*).

Figure 1 shows the air void content measured and predicted versus measured dielectric using the assumptions reported by Al-Qadi et al (*23*). The plot is adapted to convert the presented bulk specific gravity relationship to air void content which is compatible with the results of this study by using the assumed maximum specific gravity reported of 2.526. The results from the Al-Qadi study show a reasonable agreement between the measured and predicted values, although the measured data are less sensitive to changes in dielectric than the predicted response. Additionally, it should be noted that the air void contents were greater than typical air voids observed in the field because of limitations in their compaction process when preparing laboratory samples.

Whereas the Al-Qadi et al. study showed discrepancies and limitations with the laboratory calibration method and ability to match the sensitivity of measured air void content versus dielectric values, the Rayleigh model was determined to be the most rational mix model of the available choices (*23*). Although imperfect, this model can be used to quantify the expected effect of mix changes. For example, Figure 2 shows the Rayleigh model predicted air void content versus dielectric constant using the assumed values from Table 1 to compare the effect of aggregate dielectric and binder dielectric changes. The aggregate dielectric constant (Figure 2*a*) shows three aggregate value plots that represent a reasonable Minnesota source granite $e = 5$ (green), Al-Qadi et al. assumed limestone aggregate ($e = 6$), and reasonable Minnesota source limestone aggregate ($e = 7$). These represent a relatively conservative range as compared with the range of 4 to 9 for dielectric values of aggregates reported by multiple studies (*23*, *27*). The binder dielectric content (Figure 2*b*) of 2, 3, and 4 represents

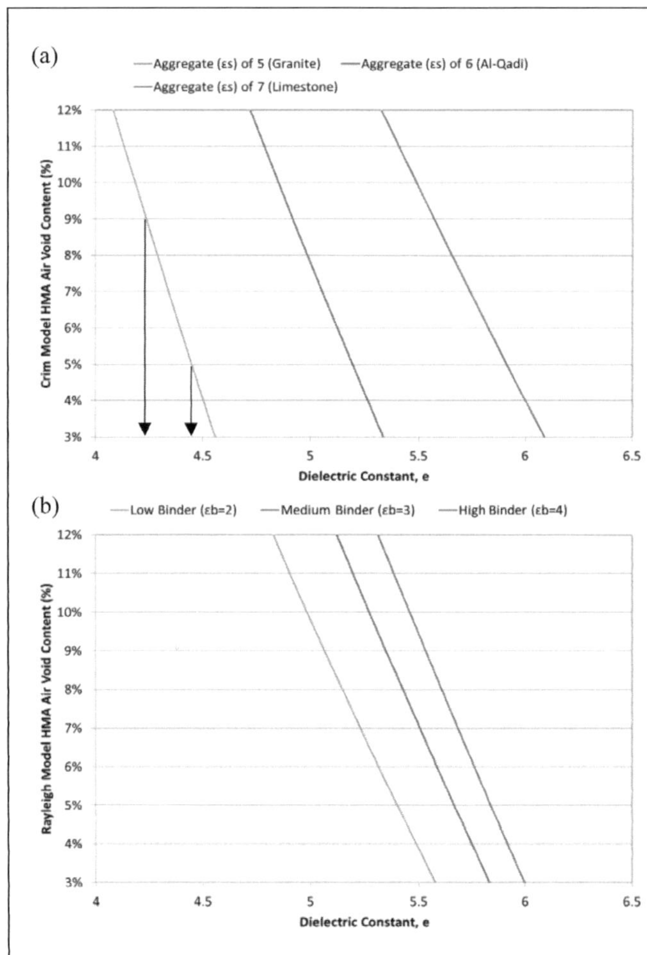

Figure 2. Predicted change in air void content using the Rayleigh model for a range in dielectric values assumed for (*a*) aggregate and (*b*) binder.

a relatively large range of dielectric values compared with the 2.5 and 3.0 values used by Araujo et al. and Al-Qadi et al., respectively (*23, 27*). It can be observed that a significantly greater effect on air void content is caused by changes in the aggregate dielectric as compared with the binder, even for relatively small changes in possible aggregate sources. For example, Araujo reported dielectric constants from 24 samples ranged from 4 to 9, including various sedimentary, magmatic, and metamorphic rocks, which matched previous studies by Ulaby et al. (*31*) and Parkhomenko (*32*). The significance of aggregate dielectric constant is a reasonable result considering the mixing rules are based on volumetric and assumed or calculated dielectric constant values of each component, and the aggregate contributes the highest percentage of the asphalt pavement volume. These types of observations are valuable in determining when the type and magnitude of mix changes may not require unique calibration and mitigate the need for additional field cores.

The significant effect of aggregate dielectric on measured dielectric has been recognized and a method of measuring the aggregate dielectric using a portable network analyzer and two cylindrical cavities to measure the dielectric constant of various aggregate sources has been developed (*27, 29, 30*). Although this approach is especially attractive in the direction of eventually eliminating the need for field cores, this method reported precision issues involving difficulties with creating a rock specimen at adequate dimensions; it also requires significant user expertise to run the test, and has to be conducted at the same frequency content as the GPR equipment used in the field. The Al-Qadi et al. and Araujo et al. studies included the Rayleigh model (least sensitive model) as the most effective model representation of asphalt pavement (*23, 27*), and this is the one used for comparison with the method proposed in this paper.

Observed Empirical Relationships versus Mix Model Predictions

Both mix model theory and observed field relationships show that the measured dielectric is a function of the type of pavement being tested in addition to the well-established relationship between air void content and dielectric constant for a given mix. Therefore, although the RDM method for assessment of compaction is valid when the pavement material is consistent, comparison of different pavement designs within a project may not be appropriate. For example, a pavement with a higher dielectric aggregate source like limestone may have a higher overall dielectric profile than a pavement with a granite aggregate source that has a low dielectric value regardless of air void content. Figure 3 gives an example of core-measured air void content versus RDM-measured dielectric constants on a pavement test-track with six different pavement mix designs at MnROAD. These measurements were made without any assumptions

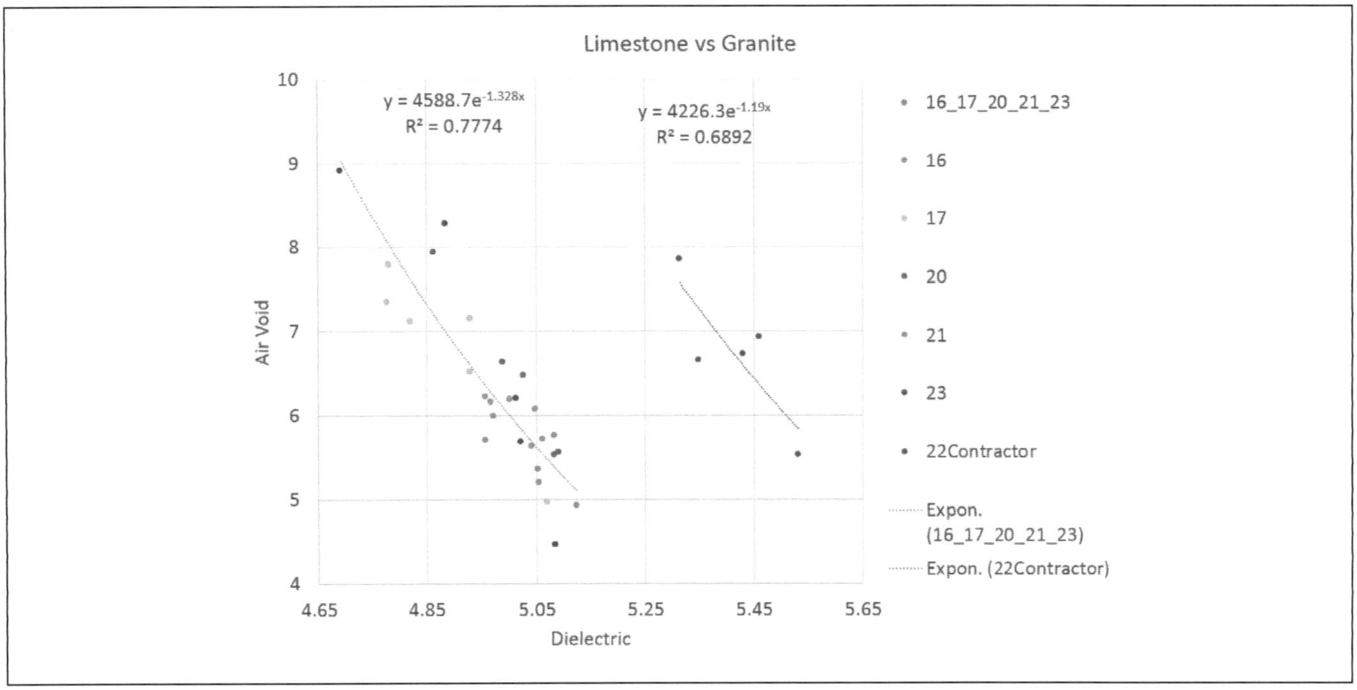

Figure 3. Example core versus dielectric data illustrating when different mixes can be combined (*left curve*) and when they need to be separated (*right curve*).

about mix components. It can be observed that a relatively good agreement between core- and RDM-measured dielectrics is obtained even when combining data from five different mix designs as long as the aggregate source is constant (granite in this case). It can also be observed that the limestone-based mix exhibited higher dielectric values than the granite-based mix regardless of the air void content. This agrees with the mix model predictions shown in Figure 2, for which changes other than aggregate dielectric constant are relatively small in comparison with the change in dielectric caused by changes in air void content.

Although the empirical data generally agree with the predicted mix model relationships, the observed sensitivity (change in air voids for a given change in dielectric) shows a similar discrepancy to that observed in the laboratory study conducted by Al-Qadi et al., even when using the Rayleigh mix model (*23*). Figure 4 shows curves developed from multiple locations (Texas, Minnesota, Maine, Nebraska), conducted by multiple agencies (University of Minnesota, Minnesota Department of Transportation, and Texas Transporatation Institute). It can be observed that each model shows a less sensitive slope between measured and mix model predicted air void content as compared with the curves shown in Figure 2. From Figure 4's superposition of an equivalent mix model curve from Figure 2 (red) along with embedded arrows in each figure placed at the granite aggregate predicted and observed curves, it can be observed that the same change from 9% air void content to 5% air void content requires over twice as much of a change in observed dielectric as the Rayleigh model predicts. This agrees with the observation from the laboratory study showing the mix models to be overly sensitive to changes in dielectric (*23*).

Proposed Model

The practicality and sensitivity issues described in the previous section preclude direct implementation of the mixing models. However, the value of incorporating a rational model in the conversion is proposed to address issues with the currently accepted conventional model. The currently accepted conventional model can be unstable in predicting air void content at the extremes (it is recommended to collect 5th and 95th percentile dielectric values to avoid extrapolation) (*18*, *19*, *21*), especially if there is a lack of dielectric versus core data to get an accurate representation. This is known from previous studies and it is recommended to select field core locations from a wide range of dielectric data to calibrate when using the currently accepted conventional method (*18*, *19*, *21*). However, selecting a wide range of core locations is not always practical. Moreover, because of a small number of replicates at the end of the measured

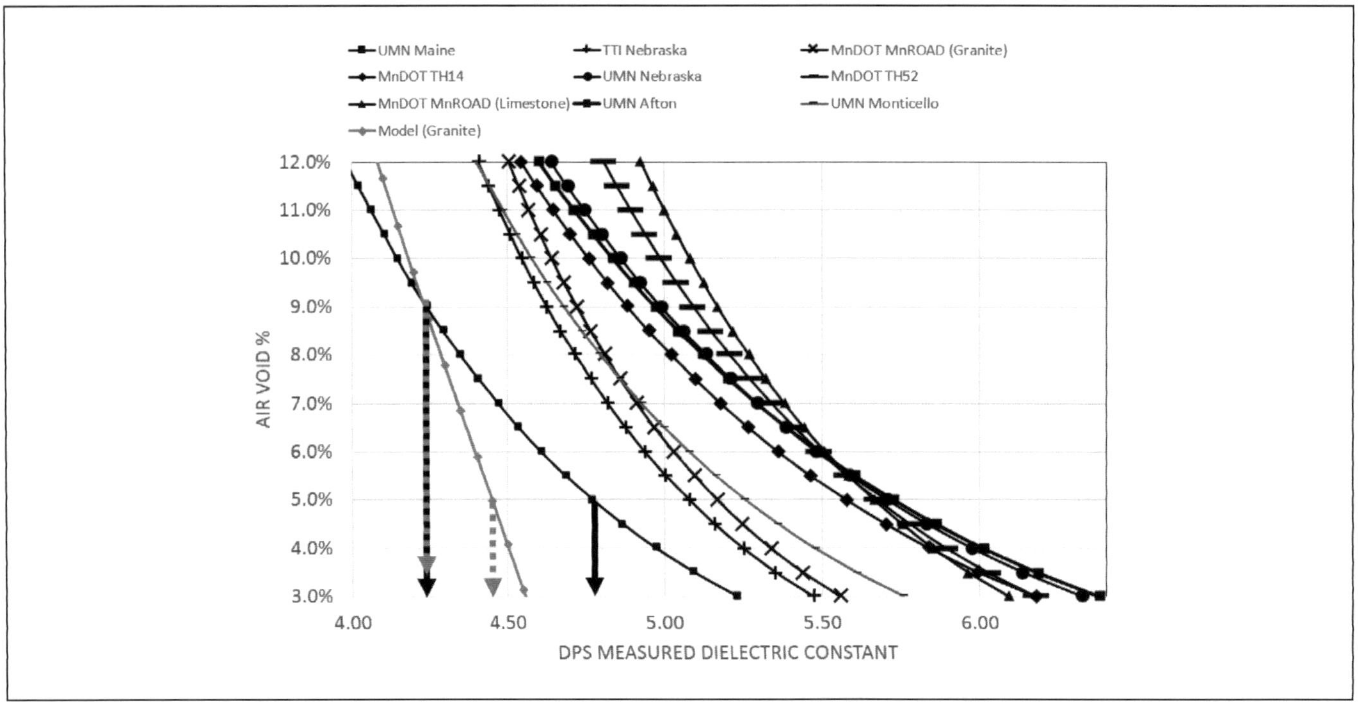

Figure 4. Relationship between dielectric and air void content for example RDM projects (*black*) in comparison with the mix model equivalent from Figure 2 (*red*).

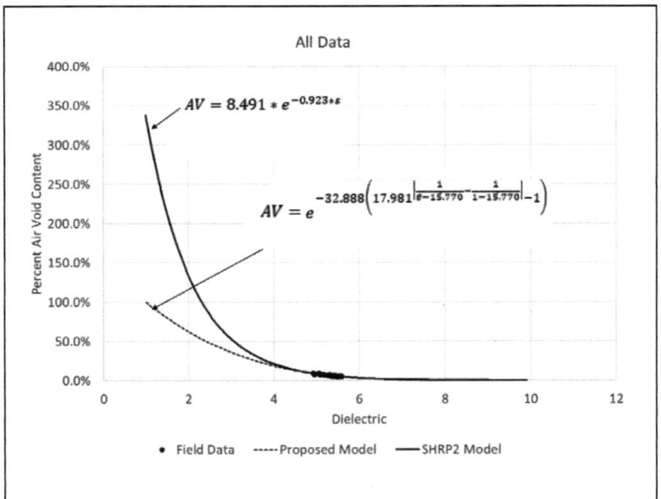

Figure 5. Field collected data along with the currently accepted and proposed models.

ranges, a small error in either the measured dielectric value at the core location or the air void value of the core may lead to a significant error in prediction, especially outside of the calibration range. This instability is magnified when extrapolating outside of the core calibrated range when using a model that is not constrained.

To evaluate the instability of the current model, consider data collected with an RDM during construction of 7 mi of 1.5-in. asphalt overlay of TH14 near Eyota, MN. Figure 5 shows 32 core-measured air void results versus RDM-measured dielectrics along with the conventional model shown in bold. It can be observed that the conventional model matches core data well. However, the conventional model predictions become unrealistic for lower dielectric values. When the dielectric constant approaches 1 a predicted air void percentage approaches 337.3%.

To address this limitation, the following model is proposed:

$$\mathrm{AV} = \exp\left(-B\left(D^{\left|\frac{1}{e-C}-\frac{1}{1-C}\right|}-1\right)\right) \tag{6}$$

where AV is the air void content, e is an input dielectric constant, and B, C, and D are constants determined by a nonlinear least-squared fit.

The proposed approach is an empirical model allowing for a slope that matches the observed field data and is constrained to physical behavior even at low dielectric/high air void content locations. If the dielectric constant input approaches 1 then the model predicts the air void content to be 100%, which is the correct value. This constraint makes the proposed model less sensitive to errors in the calibration data for cores with high air void content and more suitable for extrapolation beyond the lowest dielectric value in the calibration dataset.

The proposed model does not offer a significant improvement over the currently accepted model when many cores are available and the range of core calibrations spans the range of collected dielectric values for interpolation between known points, such as was the case in the field collected data in this study. However, the desire to limit cores, logistical challenges such as accommodating moving traffic closures, and other factors may not always be conducive to gathering a significant amount of cores covering the full measured dielectric range. These cases with extrapolation beyond the calibrated range with limited cores show the value of the proposed mechanistic model.

To demonstrate the advantage of the proposed model, both conventional and the proposed models were recalibrated using the same core data except when the measured dielectric was 5.3 or lower. Removing these highest dielectric values still provided a reasonable amount of calibration data with 12 remaining cores. To illustrate the stability of the proposed model in comparison with the currently accepted model, Figure 6 shows the predicted air void content versus a large range of dielectric values for both the conventional (solid black line) and proposed (dashed black line) models. Figure 6*a* shows the predicted air void contents with the full range of data. It can be observed from Figure 6 that within the calibration range (i.e., interpolated dielectric to air void conversions) both the conventional model and the proposed model predict similar air void contents. However the conventional model predicts a significantly higher air void content at low dielectric ranges. Comparison with Figure 5, for which the full dataset was available, shows that the calibrations with the modified dataset resulted in much greater discrepancy between the conventional and proposed model predictions than the calibrations with the original dataset. To put the discrepancy in perspective of realistically measured dielectrics for the project, Figure 6*b* shows the same comparison within a small range of dielectrics spanning the confined joint measured dielectric distribution (from 4.75 to 5.25).

It can be observed that significant discrepancies occur, even within realistic dielectric ranges with the predicted air void content from the conventional model at the low end of 4.75 resulting in a 17% air void content compared with less than 12% using the proposed method. Comparison with the full dataset predictions, expected realistic air void content, and measured core air void content support the proposed model prediction as compared with the conventional model. These observations lead to the conclusion that the proposed model offers a significant improvement of the predictions for the low dielectric value areas, especially when the accuracy of the measurements is low or an insufficient number of cores

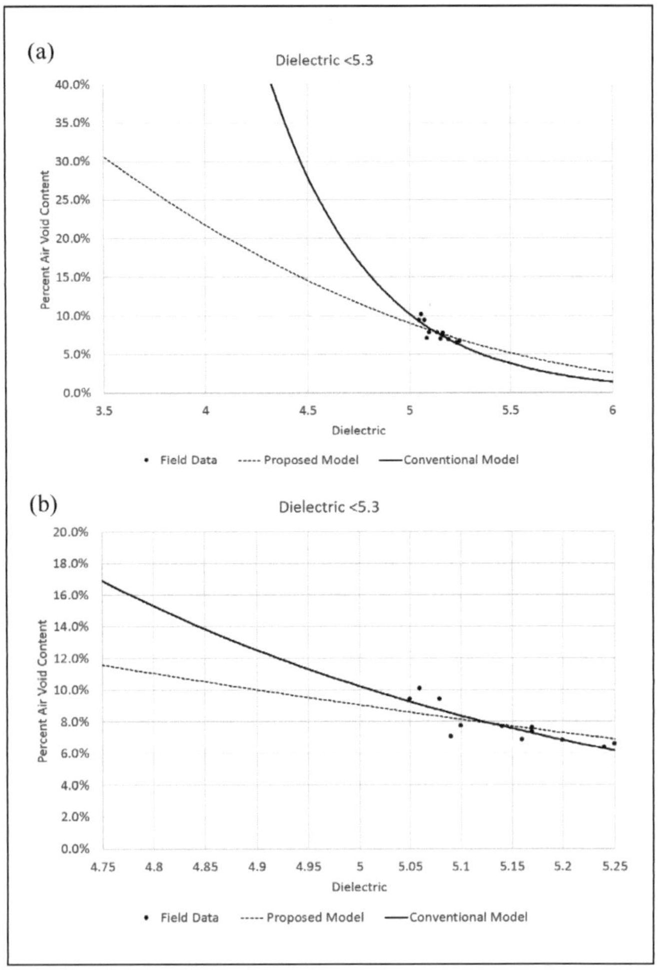

Figure 6. Comparisons of the predictions of the models calibrated with the field data and the modified dataset for (*a*) the conventional model with full range of dielectrics and (*b*) the proposed model, with realistic range of dielectrics for confined joint.

on the extremes (high or low dielectric values) is available.

Field Testing Results

As observed above, the conventional and modified models produce very similar results for a wide range of the measured dielectric values, except when the dielectric values extend to low dielectric and air void contents. Whereas the conventional model leads to unreliable and physically inadmissible air void contents for low dielectric values, the modified model offers a more rational and justifiable alternative. To evaluate the significance of this enhancement when applied to a full-scale field implementation, consider the data collected with an RDM from 7 mis of a 1.5-in. asphalt overlay of TH14 near Eyota, MN. Figure 7 shows histograms of dielectric values measured with the RDM along a stretch of pavement and separated into two groups:

- Mat—a GPR sensor path is at least 2 ft from the closest joint.
- Confined joint—a GPR sensor path is within 1 ft of the longitudinal joint compacted when the adjacent lane has been already constructed.

Using the conventional and modified calibration models, dielectric frequency distributions were converted into the distributions of the relative densities, RD, defined as:

$$RD = 1 - AV \tag{7}$$

Figure 8*a* shows the histograms obtained from these distributions with the conventional model and Figure 8*b* shows the histograms obtained with the proposed model. Table 2 presents summary statistics obtained using these

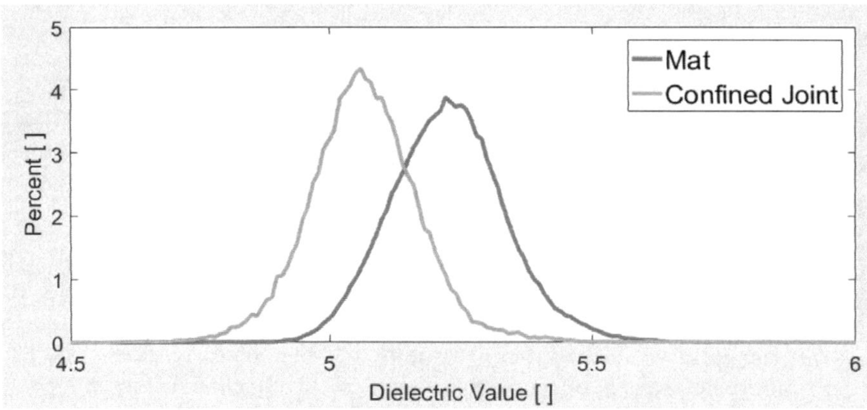

Figure 7. Distribution of dielectric value measured over a 7-mi stretch of TH14 asphalt pavement.

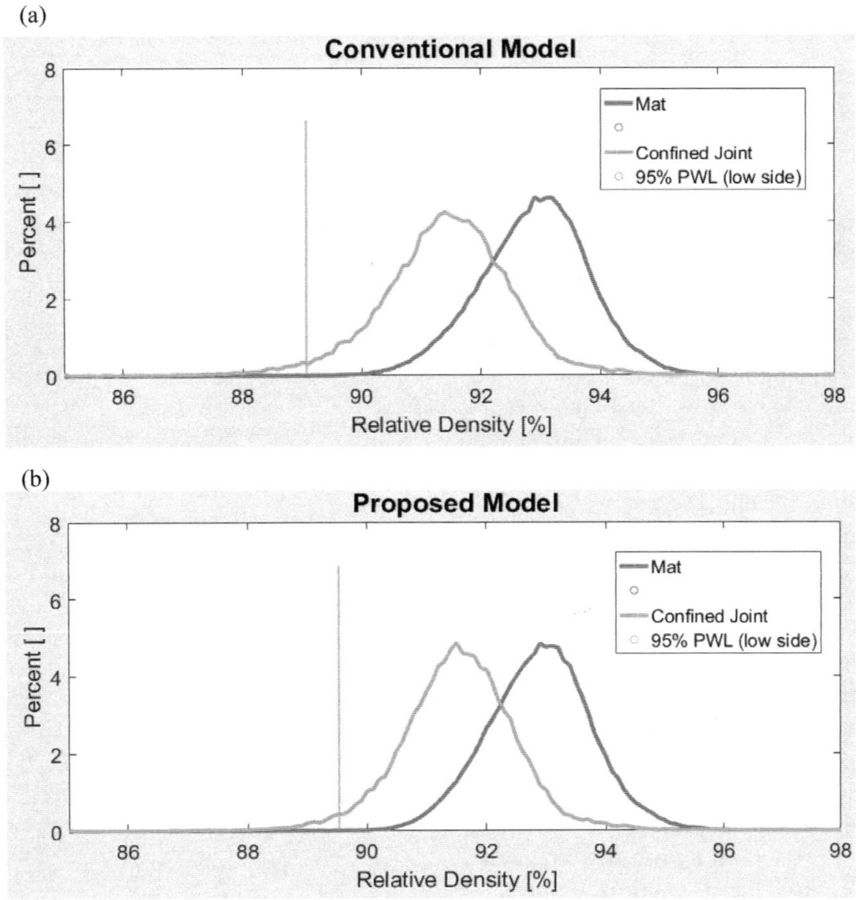

Figure 8. Histograms of the relative densities for mat and confined joint of TH14, MN, showing (*a*) the conventional model and (*b*) the proposed model.

distributions. Comparison of Figures 8*a* and 8*b* reveals that the relative density distributions for the mat are almost identical. Table 2 shows that these distributions have the same mean and median values. The relative densities corresponding to 95% within limits are also very similar: 91.0% and 91.2% from the conventional and proposed models, respectively. The confined joint exhibited a lower relative density compared with the mat compaction level, as expected. It is also not a surprise that a greater discrepancy between the relative density distributions is observed for the confined joint. Although the mean and median values are also very similar, the difference between the relative densities corresponding to 95% within limits is significantly greater: 89.0% and 89.5% from the conventional and proposed models, respectively. This suggests that the conventional model tends to overestimate the percentage of the area with low compaction. Similar analysis of the unconfined side of the joint showed an even greater difference between the models.

Conclusions

Early deterioration and long-term performance of asphalt pavements is highly affected by quality of compaction. To minimize potential delay in traffic closure on rehabilitation of heavily trafficked areas where quality of compaction is especially important, it is often desirable that the measurements can be taken over longer sections in a short timeframe immediately after final roller compaction, while still providing the necessary pavement coverage. An accurate conversion of the dielectric values to the air void content is especially important for the areas with low measured dielectric values such as longitudinal joints, because they often cause early pavement failure.

This study shows the potential of compaction assessment using a SHRP2 recommended technology referred to as the RDM, which assesses compaction using a continuous dielectric profile. A key to the success of this continuous, nondestructive technology is relating the

Table 2. Summary Statistics for Mat and Confined Joint RDM Scanned Locations

Statistical measures	Relative density			
	Conventional model		Proposed model	
	Mat	Confined joint	Mat	Confined joint
Mean	92.9%	91.5%	92.9%	91.5%
Median	92.9%	91.4%	92.9%	91.5%
Low compaction end of 95% within limits	91.0%	89.0%	91.0%	89.5%

dielectric constant to actual achieved air void content. Several approaches for development of an implementable model to convert dielectric variation to air void content variation were considered in the paper including the conventional empirical model and physics-based electromagnetic mixing models. Although the empirical model often produces reliable predictions, it requires a wide range of calibration cores to ensure stability, and produces unrealistic results for low dielectric constant values. The mixing models are advantageous in that they are derived from the dielectric properties and volume fractions of the asphalt mix components. However, it was found that the currently available mix models do not correctly predict the change in the air voids at a similar sensitivity to those observed in the field by multiple studies. A modified empirical model is introduced, matching the accuracy of prediction within the core calibration range, and improving the prediction at low air void contents. The improvement is magnified when a lack of core data and uncertainty is considered at low air void contents. The findings when using the modified empirical model as applied to a full-scale construction project are promising because of the implications in reduction of field cores necessary to convert dielectric measurements to a continuous compaction assessment of new asphalt pavement construction.

Author Contributions

The authors confirm contribution to the paper as follows: study conception and design: Kyle Hoegh, Lev Khazanovich, and Shongtao Dai; data collection: Kyle Hoegh, Trevor Steiner, and Shongtao Dai; analysis and interpretation of results: Kyle Hoegh, Shongtao Dai, Lev Khazanovich, Trevor Steiner; draft manuscript preparation: Kyle Hoegh, Shongtao Dai, Lev Khazanovich, Trevor Steiner. All authors reviewed the results and approved the final version of the manuscript.

References

1. Bonnaure, F., G. Gest, A. Gravois, and P. Uge. New Method of Predicting the Stiffness of Asphalt Paving Mixtures. *Association of Asphalt Paving Technologists Proc.*, Vol. 46, 1977, pp. 64–104.
2. Pellinen, T. K., J. Song, and S. Xiao. Characterization of Hot Mix Asphalt with Varying Air Voids Content using Triaxial Shear Strength Test. *Proc., 8th Conference on Asphalt Pavements for Southern Africa (CAPSA'04)*, Vol. 12, Sun City, South Africa, 2004.
3. Witczak, M., and O. Fonseca. Revised Predictive Model for Dynamic (Complex) Modulus of Asphalt Mixtures. *Transportation Research Record: Journal of the Transportation Research Board*, 1996. 1540: 15–23.
4. Kassem, E., A. Chowdhury, T. Scullion, and E. Masad. Measurements of Asphalt Pavement Density using Ground Penetrating Radar and Its Relationship to Performance. Presented at 91st Annual Meeting of the Transportation Research Board, Washington, D.C., 2012.
5. Linden, R. N., J. P. Mahoney, and N. C. Jackson. Effect of Compaction on Asphalt Concrete Performance. *Transportation Research Record: Journal of the Transportation Research Board*, 1989. 1217: 20–28.
6. Evans, R., M. Frost, M. Stonecliffe-Jones, and N. Dixon. A Review of Pavement Assessment using Ground Penetrating Radar (GPR). *Proc., 12th international conference on Ground Penetrating Radar*, Birmingham, UK, 2008.
7. Lai, W., T. Kind, S. Kruschwitz, J. Wöstmann, and H. Wiggenhauser. Spectral Absorption of Spatial and Temporal Ground Penetrating Radar Signals by Water in Construction Materials. *Nondestructive Testing Evaluation International*, Vol. 67, 2014, pp. 55–63.
8. Al-Qadi, I. L., D. K. Ghodgaonkar, V. K. Varada, and V. V. Varadan. Effect of Moisture on Asphaltic Concrete at Microwave Frequencies. *IEEE Transactions on Geoscience and Remote Sensing*, Vol. 29, No. 5, 1991, pp. 710–717.
9. Al-Qadi, I. L., S. Lahouar, and A. Loulizi. In Situ Measurements of Hot-Mix Asphalt Dielectric Properties. *NDT & E International*, Vol. 34, No. 6, 2001, pp. 427–434.
10. Al-Qadi, I. L., and S. Lahouar. Measuring Layer Thicknesses with GPR: Theory to Practice. *Construction and Building Materials*, Vol. 19, No. 10, 2005, pp. 763–772.
11. Leng, Z. *Prediction of In-Situ Asphalt Mixture Density using Ground Penetrating Radar: The Theoretical Development and Field Verification*. PhD thesis. University of Illinois at Urbana–Champaign, Champaign, IL, 2011.
12. Leng, Z., I. L. Al-Qadi, and S. Lahouar. Development and Validation for In Situ Asphalt Mixture Density Prediction Models. *NDT & E International*, Vol. 44, No. 4, 2011, pp. 369–375.

13. Shangguan, P., and I. L. Al-Qadi. Content-Based Image Retrieval Approaches to Interpret Ground Penetrating Radar Data. *Construction and Building Materials*, Vol. 69, 2014, pp. 10–17.
14. Shangguan, P., and I. L. Al-Qadi. Calibration of FDTD Simulation of GPR Signal for Asphalt Pavement Compaction Monitoring. *IEEE Transactions on Geoscience and Remote Sensing*, Vol. 53, No. 3, 2015, pp. 1538–1548.
15. Saarenketo, T., and T. Scullion. Road Evaluation with Ground Penetrating Radar. *Journal of Applied Geophysics*, Vol. 43, No. 2, 2000, pp. 119–138.
16. Sebesta, S., T. Scullion, and T. Saarenketo. *Using Infrared and High-Speed Ground-Penetrating Radar for Uniformity Measurements on New HMA Layers*. Report S2-R06C-RR-1. Strategic Highway Research Program, National Transportation Board, Washington, D.C., 2013.
17. Leng, Z., and I. L. Al-Qadi. An Innovative Method for Measuring Pavement Dielectric Constant using the Extended CMP Method with Two Air-Coupled GPR Systems. *NDT & E International*, Vol. 66, 2014, pp. 90–98.
18. Saarenketo, T., and P. Roimela. Ground Penetrating Radar Technique in Asphalt Pavement Density Quality Control. *Proc., 7th International Conference on Ground Penetrating Radar*, Vol. 2, 1998.
19. Popik, M., K. Maser, and C. Holzschuher. Using High-Speed Ground Penetrating Radar for Evaluation of Asphalt Density Measurements. *Annual Conference & Exhibition of the Transportation Association of Canada*, Halifax, Nova Scotia, Canada, 2010.
20. Wilson, B. T., and S. Sebesta. Comparison of Density Tests for Thin Hot-Mix Asphalt Overlays. *Transportation Research Record: Journal of the Transportation Research Board*, 2015. 2504: 148–156.
21. Hoegh, K., L. Khazanovich, S. Dai, and H. T. Yu. Evaluating Asphalt Concrete Air Void Variation via GPR Antenna Array Data. *Case Studies in Nondestructive Testing and Evaluation*, Vol. 3, 2015, pp. 27–33.
22. Maser, K., and A. Carmichael. *Ground Penetrating Radar Evaluation of New Pavement Density*, Washington Department of Transportation Report No. WA-RD 839.1, Olympia, WA, 2015. Retrieved from https://www.wsdot.wa.gov/research/reports/fullreports/839.1.pdf.
23. Al-Qadi, I., Z. Leng, S. Lahouar, and J. Baek. In-Place Hot-Mix Asphalt Density Estimation Using Ground-Penetrating Radar. *Transportation Research Record: Journal of the Transportation Research Board*, 2010. 2152: 19–27.
24. Scott, M. L., N. Gagarin, M.K. Mills, and M. Oskard. Step Frequency Ground Penetrating Radar Applications to Highway Infrastructure Measurement and System Integration Feasibility with Complementary Sensors. *AIP Conference Proceedings*, Vol. 820, No. 1, 2006, pp. 1624–1631.
25. Tong, M., and H. Tao. Permeability Estimating from Complex Resistivity Measurement of Shaly Sand Reservoir. *Geophysical Journal International*, Vol. 173, No. 2, 2008, pp. 733–739.
26. Bona, N., A. Ortenzi, and S. Capaccioli. Advances in Understanding the Relationship between Rock Wettability and High-Frequency Dielectric Response. *Journal of Petroleum Science and Engineering*, Vol. 33, No. 1, 2002, pp. 87–99.
27. Araujo, S., L. Delbreilh, L. Laguerre, H. Dumont, E. Dargent, and C. Fauchard. Rock Permittivity Characterization and Application of Electromagnetic Mixing Models for Density/Compactness Assessment of HMA by Means of Step-Frequency Radar. *Near Surface Geophysics*, Vol. 14, No. 6, 2016, pp. 551–562.
28. Benedetto, A., and F. Tosti. Inferring Bearing Ratio of Unbound Materials from Dielectric Properties using GPR: The Case of Runaway Safety Areas. *Airfield and Highway Pavement 2013: Sustainable and Efficient Pavements*. 2013, pp. 1336–1347.
29. Fauchard, C., B. Beaucamp, and L. Laguerre. Non-Destructive Assessment of Hot Mix Asphalt Compaction/Density with a Step-Frequency Radar: Case Study on a Newly Paved Road. *Near Surface Geophysics*, Vol. 13, No. 3, 2015, pp. 289–297.
30. Fauchard, C., B. Li, L. Laguerre, B. Héritier, N. Benjelloun, and M. Kadi. Determination of the Compaction of Hot Mix Asphalt using High-Frequency Electromagnetic Methods. *NDT & E International*, Vol. 60, 2013, pp. 40–51.
31. Ulaby, F. T., T. H. Bengal, M. C. Dobson, J. R. East, J. B. Garvin, and D. L. Evans. Microwave Dielectric Properties of Dry Rocks. *IEEE Transactions on Geoscience and Remote Sensing*, Vol. 28, No. 3, 1990, pp. 325–336.
32. Parkhomenko, E. I. *Electrical Properties of Rocks*. Springer Science & Business Media, Berlin, Germany, 2012.

The Standing Committee on Asphalt Pavement Construction and Rehabilitation (AFH60) peer-reviewed this paper (18-06229).

Article

TRR
JOURNAL OF THE TRANSPORTATION RESEARCH BOARD

Transportation Research Record
2018, Vol. 2672(26) 155–164

DOI: 10.1177/0361198118791916
journals.sagepub.com/home/trr
SAGE

Evaluation of Mix Designs for Rapid Construction of Bonded Concrete Overlays on Asphalt

Julio Paniagua Fernandez[1], Fabian Paniagua Fernandez[1], John Harvey[1], and Angel Mateos[2]

Abstract

Road users in California have a low tolerance for delays during pavement rehabilitation. Therefore, the California Department of Transportation (Caltrans) is interested in evaluating bonded concrete overlays on asphalt (BCOA) as a potential rehabilitation alternative for distressed asphalt pavement. The use of high early strength concrete to facilitate rapid construction of BCOA has not been used in other U.S. states. This paper summarizes a complete study for the evaluation of concrete mix designs for rapid construction of BCOA that were placed on a set of full-scale test sections for accelerated pavement testing. Two types of mixes were created according to selected targets for opening times (OT) to traffic. The concrete mixes for use in night closures require an opening time of four hours, and the mixes for use in 55-hour weekend closures require an OT of 10 hours. Type III Portland cement and calcium sulfoaluminate (CSA) cement were used to design the 4H-OT mixes. Two 10H-OT mixes were designed using the same Type II/V portland cement. One of the 10H-OT mixes included light weight aggregate instead of a portion of conventional sand to evaluate the effects of this approach for internal curing to reduce drying shrinkage stresses.

Bonded Concrete Overlay on Asphalt (BCOA) is currently defined in an evaluation and development study underway in California as jointed plain concrete overlay placed on existing asphalt pavement with slab thicknesses between 100 and 175 mm (4–7 in.). The bonding of the slabs to the underlying asphalt layer is critical so that they behave as a monolithic layer to resist bending. This moves the neutral axis down in the combined layer compared with the middle of the slab if unbonded and thus reduces tensile stresses in the slab (*1*). Long-term performance is also aided by using short joint spacings to reduce curling and warping stresses caused by temperature and drying shrinkage gradients and bending stresses produced by applied loads (*1*). Slab sizes being looked at include full lane width 3.7 by 3.7 m (12 by 12 Ft), 1.8 by 1.8 m (6 by 6 Ft), and 2.4 by 2.4 m (8 by 8 Ft) for use on the outside of the outer lane with the extra width used to keep traffic off the slab edge.

Even though BCOA is a rehabilitation technique that has been developed and used in different U.S. states (*2*) as well as in other countries in the Americas, Europe, and Asia, Caltrans is interested in further developing BCOA methodologies for rapid construction. Existing design methodologies, such as the bonded concrete overlay of asphalt pavements mechanistic-empirical design guide (BCOA-ME) (*3*) prepared by the University of Pittsburgh are being used as a starting point for further development. The rapid construction aims to use night and weekend closures with time to traffic opening after placement of the concrete of 4–12 hours. To achieve these strengths in such a short period, high early strength concrete (HESC) needs to be used.

Project Background

The work done at the University of California Pavement Research Center (UCPRC) presented in this paper is part of a California Department of Transportation (Caltrans) sponsored project to advance current BCOA knowledge developed in other states. In particular, this

[1]Department of Civil and Environmental Engineering, University of California Pavement Research Center, University of California, Davis, CA
[2]University of California Pavement Research Center, University of California, Berkeley, Richmond, CA

Corresponding Author:
Address correspondence to Julio Paniagua Fernandez: jpaniaguaf@ucdavis.edu

Table 1. Full-Scale Test Factorial

ID	Slab dim. (m × m)	Slab thickness (mm)	HESC mix	Asphalt surface treatment	Asphalt base type	Asphalt base thickness (mm)
A	1.8 × 1.8	150	Type II/V-10H OT	Micromilling	Old HMA	120
B	1.8 × 1.8	115	Type II/V-10H OT	Micromilling	Old HMA	120
C	1.8 × 1.8	115	Type II/V-10H OT	Milling	Old HMA	120
D	2.4 × 2.4	115	Type II/V-10H OT	Micromilling	Old HMA	60
E	1.8 × 1.8	115	Type II/V-10H OT	Micromilling	Old HMA	60
F	1.8 × 1.8	115	Type II/V-10H OT	Micromilling	New RHMA-G	120
G	1.8 × 1.8	115	Type III-4H OT	Sweeping	New RHMA-G	120
H	1.8 × 1.8	115	CSA-4H OT	Sweeping	New RHMA-G	120
I	1.8 × 1.8	115	Type II/V-10H OT	Sweeping	New RHMA-G	120
J	3.6 × 3.6	115	Type II/V-10H OT	Sweeping	New RHMA-G	120
K	3.6 × 3.6	115	Type II/V LWA-10H OT	Sweeping	New RHMA-G	120

study is considering California's dryer climate conditions and Caltrans' need for fast construction. It is expected that BCOA will need to be constructed primarily in 10-hour night time and 55-hour weekend traffic closures to be competitive with asphalt overlays from a traffic congestion perspective. The goal of the overall project research is to develop recommendations and guidance on the use of BCOA as a rehabilitation alternative in California. This paper focuses on the laboratory performance evaluation of the high early strength concrete (HESC) mixes to be used for BCOA test sections for full-scale accelerated pavement testing (APT).

The APT was performed by two Heavy Vehicle Simulators (HVS) on 11 sections that were built at the UCPRC Davis facility. The 11 sections (Table 1) follow a factorial design in which the variables are: HESC mix type; slab thickness; asphalt concrete surface type, condition and treatment; and asphalt base thickness.

The concrete mixes that are being used in this project meet 2015 Caltrans Standard Specifications Section 40 "Concrete pavement" (*4*), and 2010 Standard Special Provisions 40-5_A11-15-13, which cover construction of jointed plain concrete pavements with Rapid Strength Concrete (RSC). The RSC specification requires a minimum flexural strength of 2.8 MPa (400 psi) for concrete slabs before opening to highway traffic (*5*). The concrete mixes for the night closure required an opening time (OT) to traffic of four hours for which the acronym 4H-OT is used in this paper. The mixes for the weekend closure require an OT to traffic of 10 hours (10H-OT). Two mixes were designed for each of the two opening time targets. Portland cement Type III and calcium sulfoaluminate (CSA) were used to design the 4H-OT mixes (Type III-4H-OT, CSA-4H-OT, respectively). Both of the 10H-OT mixes used the same portland cement Type II/V. Caltrans has required use of Type II or V for many years due to the widespread presence of sulfate-rich soils. The difference between the two 10H-OT mixes is the use of Light Weight Aggregate (LWA) in place of a portion of conventional sand. The goal of the LWA was to reduce internal drying vertical shrinkage gradients in the slabs which can cause very high tensile stresses, particularly in the dry climates typical of most of California. These mixes are identified as Type II/V-10H-OT and Type II/V LWA-10H-OT.

BCOA is intended to be placed on top of distressed asphalt pavement. For this reason, a set of test sections were built on top of a previously tested full-depth reclamation structure. The underlying structure a 120 mm (4.7 in.) asphalt pavement and 130 mm (5.1 in.) Class 2 aggregate base. This structure was constructed 3.5 years before BCOA construction. The asphalt pavement, which is identified as "old HMA," was a dense graded hot mix asphalt, 19 mm (3/4 in.) nominal maximum aggregate size (NMAS), with PG 64-16 binder and 15% reclaimed asphalt pavement (RAP). The condition of these flexible sections before BCOA construction was relatively poor, mainly due to permanent deformation. Some localized areas had 20–25 mm (3/4–1 in.) rutting and/or fatigue cracking.

Existing literature has recommended that BCOA should be placed on aged asphalt pavement in fair to good surface condition (*6*). The placement of a new thin asphalt surface as the base for BCOA is also considered in this project to expand the range of applicability to pavements in relatively poor condition. Caltrans is interested in the evaluation of rubberized mixes for this purpose. To evaluate the use of new asphalt bases under BCOA, a new asphalt structure was paved five months before BCOA construction. A 90 mm (3.5 in.) thick dense graded hot mix asphalt (HMA) layer was placed, followed by a 30 mm (1.2 in.) layer of rubberized mix. The HMA was 19 mm (3/4 in.) NMAS, with PG 64-16 binder, and 15% RAP. The rubberized mix was gap graded

(RHMA-G), 12.5 mm (1/2 in.) NMAS, and 7.5% asphalt rubber binder content. Asphalt rubber binder included PG 70-10 base binder and 19.5% crumb rubber modifier. This base structure was called "New RHMA-G".

Goals and Objectives

The goal of the project is to develop recommendations and guidance for use of BCOA as a rehabilitation alternative in California. The goal of this paper is to present development, characterization, and performance of HESC mixes for rapid construction of BCOA that were used in the subsequent APT phase of the project.

The main objective of the concrete mix design phase of the project was to evaluate the laboratory performance of HESC mixes designed for the different early opening times and to analyze trade-offs between opening time and performance related properties. The questions to be answered were:

- What were the main results that occurred through the optimization of the mixes?
- What are the differences in main properties and contributions of each cementitious material?
- Are Caltrans strength minimum requirements very different from the minimum requirements of other agencies?
- What are the pros and cons of using LWA instead of sand?
- Can flexural strength be predicted by using compressive strength data for the mixes considered for rapid construction of BCOA?
- Do maturity readings do a good job of predicting compressive and flexural concrete strengths for the four mixes?
- How will the four mixes rank according to free relative shrinkage and drying shrinkage?
- Which are the differences and trade-offs within mixes for short- and long-term performance?

Experiment Design

Mix Designs

The targets for OT to traffic of the concrete mixes were established by Caltrans to minimize construction-related traffic congestion. Mix designs were provided by the concrete industry to facilitate future implementation without major modifications. The mix designs were verified in the UCPRC laboratory with industry and Caltrans oversight to be sure that techniques for working with these special mixes and the results were replicable, and then used for test section construction. Portland cement based designs were supplied by Teichert Materials and the CSA mix design was supplied by CTS Cement. All the cement mixes fulfill ASTM C150 requirements. The four mixes used the same alluvial aggregate from Teichert's Perkins plant.

Different admixtures were used in the four mixes and included an accelerator, two types of high-range water reducer admixture (HRWR) and a stabilizer. Accelerating admixtures are materials that aid early strength development and reduce the set time of the mixture (*7*). The high-range water reducer admixtures, also known as superplasticizers, help to achieve higher strength by decreasing the water/cement ratio at the same workability (*8*). The stabilizer (retarder) is required first, to slow the hydration process of cementing materials and then allow activation and reestablishment of the normal hydration process (*9*).

Several trials were done with each mix using flexural strength evolution results to show that the same results could be obtained in the UCPRC laboratory that were obtained by Teichert and CTS Rapid Cement in full-scale production. Industry oversight showed that the main source of difference in the initial trials was temperature control of the materials during mixing.

Test Methods and Procedures

The bullet list below shows the standard tests and procedures used in this research, with the exception noted below. The UCPRC staff involved in mix production, specimen fabrication, and specimens testing were ACI certified to ensure high quality in sampling, production, and testing of concrete mixtures.

- Concrete specimen fabrication and curing: ASTM C 192, ASTM C31
- Compressive Strength: ASTM C 39, ASTM C1231
- Flexural Strength: ASTM C 78
- Shrinkage of concrete: ASTM C 157
- Concrete setting time: ASTM C 403
- Maturity method: ASTM C 1074
- Concrete modulus of elasticity: ASTM C 469
- Coefficient of thermal expansion: AASTHO T336
- Absorption of LWA: ASTM 1761-13b
- Sampling freshly mixed concrete: ASTM C 172
- Slump of freshly mixed concrete: ASTM C 143
- Temperature of freshly mixed concrete: ASTM C 1064

An alternative to the ASTM C 192 curing procedure was developed to more closely simulate the conditions in the field and to boost the hydration and heat development. The standard curing entails free water maintained on the entire specimen surface area at all times at 23°C (73.4°F). This curing procedure might not fully represent standard

BCOA conditions for rapid construction in California. Specimens were cast over a plywood sheet to avoid heat transfer between the specimen and the ground and then placed inside a foam box. Given that the laboratory facility is not temperature controlled, temperature monitoring was needed to ensure consistent understanding of conditioning of the specimens using the principle of maturity. After two–three hours the box was located inside a building to avoid cold temperatures and wind affects. For tests before 24 hours, the specimens were removed from the box, unmolded and tested. Otherwise, the specimens stayed in the box for 24 hours and were then unmolded and placed in a water bath at a controlled temperature (25°C) until testing.

Mix Design Constraints

According to industry experts, the mixing temperature should start around 21°C (70°F) and the concrete temperature when discharged should be around 27°C (80°F). Mixing temperature was very difficult to control because of daily temperature fluctuations, especially when temperatures dropped below 10°C (50°F) in winter. To optimize and validate the mix designs, several batch trials were produced to provide feedback and knowledge for the overall project. Portland cement type III mix produces very fast reactions, but not as fast as the CSA mix reactions. After temperature control, segregation and stickiness were the other main issues faced during the validation process. It was found to be mandatory to maintain the temperature above 21°C (70°F) during mixing and discharge above 24°C (75°F) to consistently meet the threshold of 2.8 MPa (400 psi) at four hours.

The CSA mix for 4H-OT was easier to validate because the factors and consequences were more predictable and easy to understand. It is clear that it undergoes the fastest reaction process which starts immediately after the cement is in contact with the water, and in consequence mixing and specimen fabrication needs to be done precisely. This concrete mix was not as sensitive to low mixing temperature as the portland cement mixes, but high mixing temperatures need to be avoided as well because its fast reaction process could prevent the fabrication of specimens.

At the beginning it was very difficult to achieve minimum strengths with the Type II/V concrete mix. Temperature and clumping were two of the issues faced with this mix, and additional trial batches along with mix adjustments were needed to get to the strength goal.

Overall the major issue in achieving strength requirements with the HESC mixes is control of the temperature of the fresh concrete. Temperature control needs to be done before, during and after mixing. From industry advice and later from trial mixes, this research group can determine the importance of mixing within the range of 24 to 27°C (75–80°F). The ambient conditions change daily, therefore strict control of the temperatures in the aggregates, mixer drum, and water is needed to achieve the desired mixing temperature.

Final Mix Designs

Industry experts and the research team met after several trial batches whose results were not consistently satisfactory. Mixing time, admixture dosage, and heat development during curing were evaluated and improved. The mixing time of concrete depends mainly on the distance between the plant and the job site which would normally be above 30 minutes. Teichert specialists believed that these HESC mixes, when mixed in the laboratory, need prolonged mixing times so that reactions are accelerated. This was not being replicated in the laboratory due to equipment restrictions and they were only being mixed for 3–5 minutes after all the materials were added. The adjustments made were to increase the mixing period for both Portland cement mixtures and to not add the stabilizer admixture (Delvo™) in the Type III mix.

Mixing the Delvo® for such a short period was counterproductive because of the retarding effect of the admixture. The bad side of not adding this admixture was the effect on the workability which was reflected in the sinking of aggregates and stickiness. More accelerator admixture was added to the 10H-Type II/V concrete mix to accelerate the strength development. The foam curing boxes helped to control heat production during curing.

The following table (Table 2) shows the final designs for the HESC mixes, including the mix design for the 10H opening time mix using LWA. The step-by-step for the final procedure can be found in the complete report of the concrete mix design phase of this research project (*10*).

Laboratory Performance Analysis

Cementitious Materials

The different compounds of the cement play an important role in the performance of the concrete for both strength gain and shrinkage (*9*). Understanding these roles helps in design to increase the mix performance and in making the decision as to which cement to use. The silicates are the reactive compounds of the cement and determine how quickly a concrete mix will set (*11*). The tricalcium and dicalcium silicate compositions of both portland cement mixes (Type II/V and Type III) are very similar. These components control the early and final strength of the mix. Nevertheless, the component C_3A (tricalcium aluminate) has a very important role in early strength gain as well, as does the grain size of the cement

Table 2. Final HESC Mix Designs

Mix	10H-Type II/V		4H-Type III		4H-CSA		10H-Type II/V LWA	
Material	Qty.	Units	Qty.	Units	Qty.	Units	Qty.	Units
Cement	362.4	Kg	362.4	Kg	309.3	Kg	362.4	Kg
Water	121.1	Kg	113.4	Kg	130.2	Kg	121.1	Kg
CA1: Perkins 1"× #4	635.0	Kg	635	Kg	816	Kg	635	Kg
CA2: Perkings 3/8"× #8	249.5	Kg	249.5	Kg	0	Kg	249.5	Kg
FA1: Perkins Concrete Sand	518.4	Kg	538.8	Kg	571.5	Kg	174.6	Kg
FA2: Light Weight Agg. (LWA)	0	Kg	0	Kg	0	Kg	191	Kg
MasterGlenium® 3400	0	Liter	0	Liter	1.61	Liter	0	Liter
MasterGlenium® 7500	2.83	Liter	2.95	Liter	0	Liter	2.83	Liter
MasterSet® AC 534	17.71	Liter	17.71	Liter	0	Liter	17.71	Liter
MasterSet® Delvo™	0	Liter	10.40	Liter	2.02	Liter	0	Liter
Designed unit weight	2470	Kg/m^3	2482	Kg/m^3	2389	Kg/m^3	2314	Kg/m^3
Designed w/c ratio.	0.33	–	0.31	–	0.42	–	0.33	–
Designed volume	0.764	m^3	0.764	m^3	0.764	m^3	0.764	m^3
Specified slump	15-25	cm	18-28	cm	10-20	cm	15-25	cm
Set time	75	Min	61	Min	29	Min	16	Min

Note: Mass quantities per cubic meter of concrete.

particles. Flexural strength results at Opening Time and at 45 days for both portland cement mixes (Type II/V and Type III) show very similar results, with OT for the Type III mix being earlier (Figure 1).

CSA cement has a different chemical composition that does not include the C_3S as the reactive component but instead includes the C_4A_3S component to boost the early strength. The strength results at OT were almost identical to the results obtained for the portland cement mixes, nevertheless, the preliminary results showed that the CSA cement was reaching the OT strength requirement of four hours within just two hours of pouring the specimens. Long-term strength was reached early with little additional change at 45 days. The CSA flexural strength at 45 days is about 50% of the flexural strength of the portland cement mixes as can be seen in Figure 1. The other notable performance characteristic of the CSA concrete mix is its low shrinkage when reactions are occurring (*12*). This effect is due to an exothermic and expansion reaction that takes place in the C_4A_3S during the hydration process. The results in Figure 2 show the total shrinkage measured in independent relative-free shrinkage beams that were cast during construction and set on the side of the test track. The fluctuation of the shrinkage occurred as a direct response to rain events, which are plotted in light blue. Every rain event reversed a portion of the shrinkage at each point. BCOA structural design needs to consider both of these performance related properties (strength and shrinkage).

Caltrans Strength Minimum Requirements

Caltrans stipulates in section 40-5.01D(8) of the Standard Specifications (*5*) that any high-early strength concrete should reach at least 2.8 MPa (400 psi) before it is opened to traffic. This strength threshold is applied to all potential applications including BCOA. It was important to validate the Caltrans threshold by comparing it with other DOT requirements. A survey was performed within eight U.S. states (CA, MN, CO, FL, IA, WA, OH, MO) that have worked with BCOA and also have HESC strength requirements. Half of these specifications correspond to compressive testing, and Minnesota specifies both compressive and flexural strengths. To compare Caltrans specifications, all the compressive strengths were converted to flexural strengths using the equation recommended by the American Concrete Institution (ACI) Committee 330 (*13*). The relationship, originally published by Raphael (*14*) has proved to be statistically significant in predicting flexural strength from compressive strength (*15*). The minimum and maximum values that are specified by other states are 2.7 MPa (389 psi) and 3.2 MPa (450 psi). The average value among all the specified values is 2.95 MPa (427 psi) which almost matches the Caltrans strength requirement. The survey results indicate that the Caltrans minimum specification for opening strength is not different from other DOT specifications for HESC.

Light Weight Aggregate Benefits

Light weight aggregate (LWA) was tried to provide internal curing to reduce drying shrinkage stresses. The common feature of lightweight aggregates is their high internal porosity, which is the prime reason for their low bulk specific gravity, and that allows better moisture storage inside the concrete to replace water lost due to chemical shrinkage (hydration) (*3*). LWA also helps to

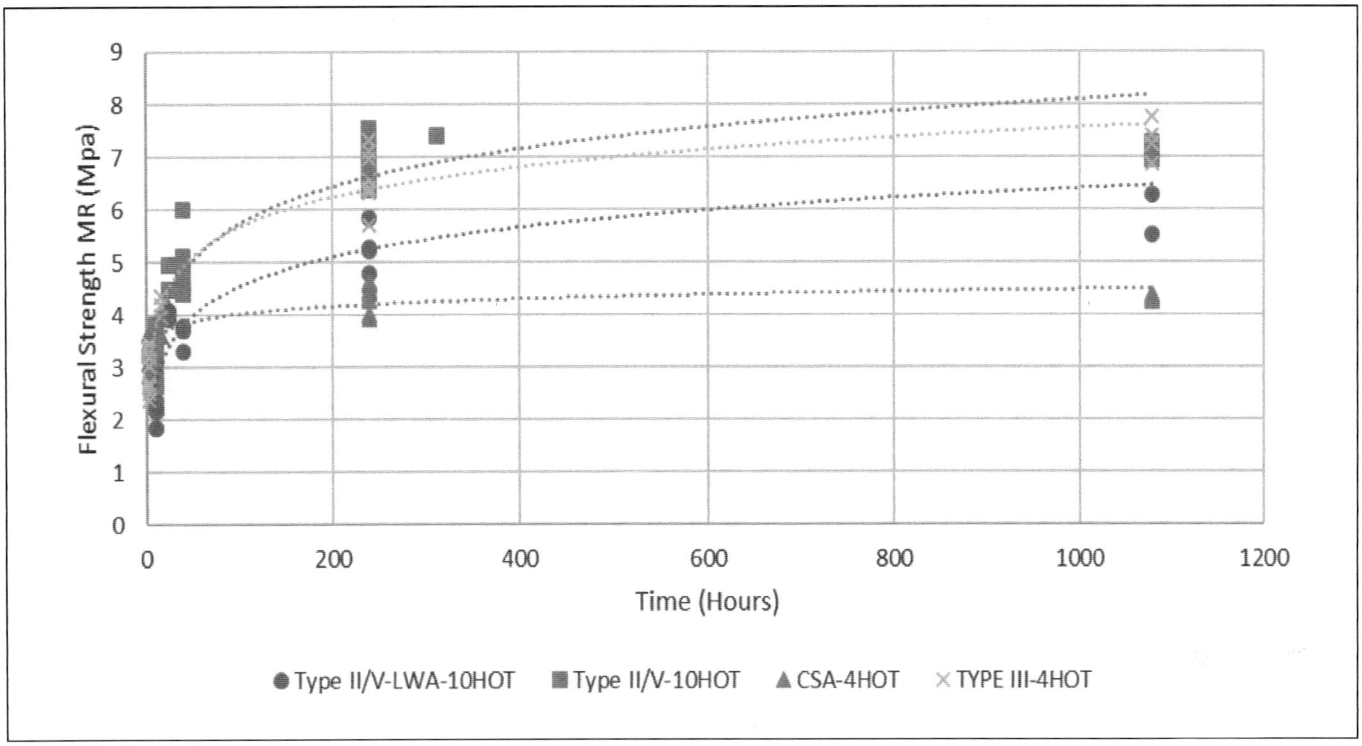

Figure 1. Flexural strength vs. time for all specimens for all HESC mixes.

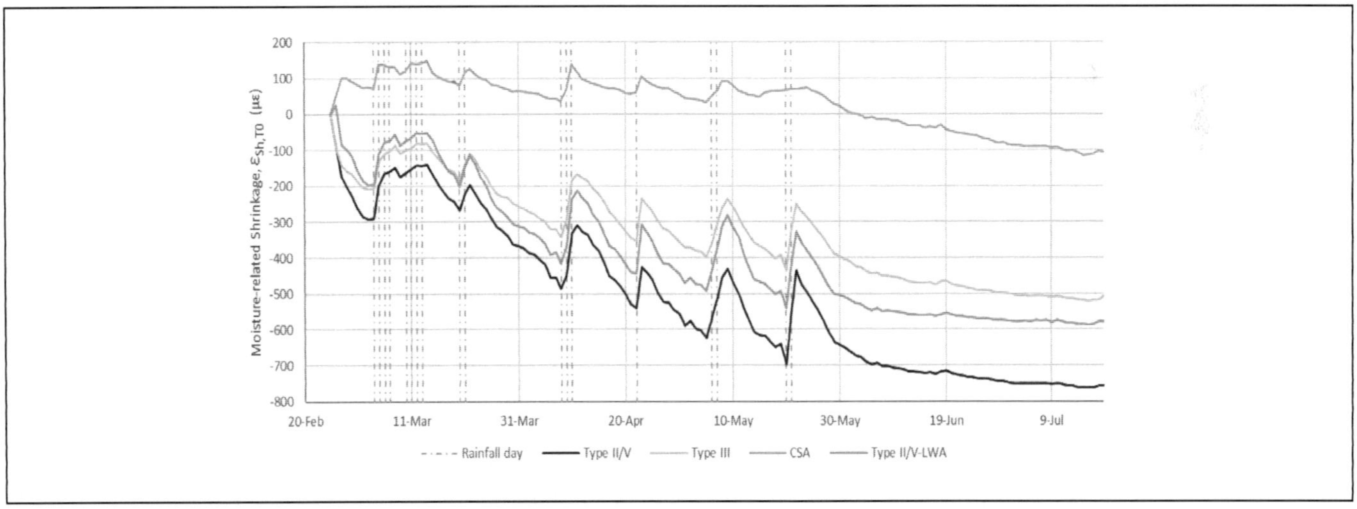

Figure 2. Shrinkage results for all HESC mixes for BCOA.

improve the autogenous shrinkage performance in the concrete, as can be seen in Figure 2 where the Type II/V LWA trend line stays above the reference mix between 12 and 24 hours after the beam was cast. Concrete including LWA develops lower a modulus of elasticity (MOE) which potentially decreases the risk of cracking. At 45 days, the LWA specimens average MOE was 12% less than the average MOE of the specimens without LWA. On the other hand, concrete with LWA does not develop as high flexural strengths as the concrete without LWA (Figure 1). Specimens with 50.7% substitution of concrete sand with LWA had an average 12.5% lower flexural strength than the results of the specimens without LWA. Results showed one-third less final shrinkage in the Type II/V concrete mixes with LWA compared with those without LWA (Figure 2).

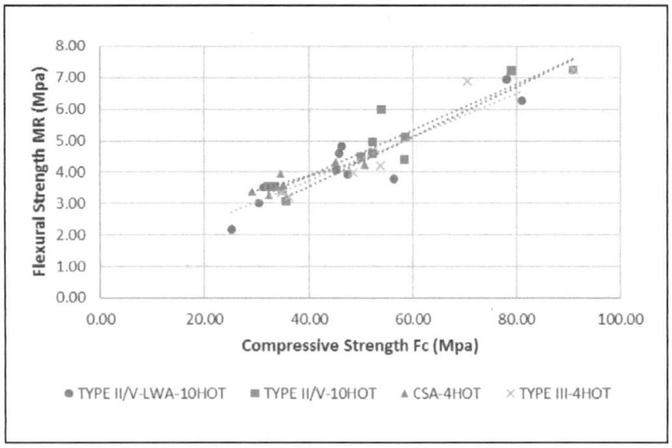

Figure 3. Flexural to compressive strength ratios for All HESC mixes.

Flexural Strength Prediction

Specimen preparation, curing, transportation, and testing are easier for compressive strength testing than for flexural beams, and relationships for using compressive strength to estimate flexural strength for HESC mixes were explored.

Flexural strength predictive equations were developed for each of the concrete mixes, and one more for the complete set of results (regardless of the concrete mix). All of them show, within a 95% confidence level, that they are statistically significant. The data shows that at an early age the flexural strength is around 10% of the compressive strength, while longer term strengths (45 days) have a ratio of about 8%. Figure 3 shows graphically the relationship for the four mixes, and the similar behavior at the different curing times: opening time, four times opening time and 45 days.

Maturity Approach

Maturity is defined as the extent of cement hydration in a given concrete mixture which causes concrete strength to be directly related to both age and its temperature history. The maturity index of concrete is used to estimate its strength at a specific time based on a pre-determined calibration of the time – temperature – strength relationship developed from laboratory tests. Even though the maturity relationships developed in this project were statistically significant, the method would not be valuable if it did not explain better the strength gain phenomenon. A prediction model depending on only time and not temperature was created for each of the four mixes and the statistical parameters obtained in the time-strength relationship were very similar to the maturity (time-temperature dependent) model parameters. Thus, for the curing conditions used here, the maturity models do not improve the predictions obtained with only a time-strength relationship.

Shrinkage Performance

The drying shrinkage test differs from the standard ASTM C157 with regard to the curing procedure. ASTM C 157 is intended for conventional portland cement concrete which has longer curing times and does not resemble the curing period of HESC mixtures. Not only is the curing time long (28 days) but also the collection of data starts many days after casting the concrete. The concrete mixtures being used in this research have hydration processes that occur within a few hours and it is important to capture their early behavior. The specimens were maintained in a 100% relative humidity room until being demolded, then submerged in lime water just after the OT (either 4 or 10 hours) for three days and afterwards they were stored in a drying room. Length measurements started when the specimens were removed from demolded.

The analysis showed that CSA-4H-OT mix performs better in shrinkage, and measurements show an expansion happening through time. On the other hand, portland cement mixes with Type II/V cement show the highest values of shrinkage. Three different analyses were performed: the standardized drying shrinkage test (as explained above), relatively-free shrinkage beams that were cast during construction of the BCOA sections and mean shrinkage measurements obtained from full-size environmental test section slabs. The relatively-free shrinkage beams were $50 \times 15 \times 4\,cm^3$ ($19.6 \times 5.9 \times 1.6\,in.^3$) and were cast during test track construction. The beams carry a vibration strain wire in their middle to record any contraction due to shrinkage. The displacements measured in these beams belong to the total shrinkage that the specimen is undergoing.

The results of the standard drying shrinkage tests became unreliable because of inconsistency of the early measurements such as abrupt expansion at initial stages. The reason for these unexpected results is attributed to measurement errors and high variability from the measurement apparatus. One length comparator has been used as well as one reference bar. On the other hand, the trends on the free shrinkage beams results appear to be more reliable and consistent, although they are not a realistic representation of the constrained full-scale slabs. The ranking in order from highest to lowest shrinkage is Type II/V, Type II/V LWA, Type III and CSA. The only disagreement with expected behavior is that Type III was expected to have higher shrinkage. Type III beams stayed on the bottom side of the mold for 24 hours more due to human error. After taking demolding it could be observed that the beams were very wet on the bottom

Table 3. Flexural Strengths for Specimens Prepared in the Laboratory for All HESC Mixes

Concrete mix	Test	Age		MR (psi)	MR (Mpa)	COV
TYPE III-4H OT	MR-OT	4	Hours	482.84	3.33	3.2
	MR-4OT	16	Hours	606.05	4.18	3.7
	MR-45D	45	Days	1060.63	7.31	4.3
CSA-4H OT	MR-OT	4	Hours	486.94	3.36	2.4
	MR-4OT	16	Hours	559.58	3.86	4.7
	MR-45D	45	Days	624.98	4.31	1.3
TYPE II/V-10H OT	MR-OT	10	Hours	485.05	3.34	5.7
	MR-24H	24	Hours	685.18	4.72	4.8
	MR-4OT	40	Hours	725.32	5.00	11.0
	MR-45D	45	Days	1035.23	7.14	2.0
TYPE II/V-LWA-10H OT	MR-OT	10	Hours	406.49	2.80	24.6
	MR-24H	24	Hours	579.63	4.00	1.9
	MR-4OT	40	Hours	585.90	4.04	14.2
	MR-45D	45	Days	905.84	6.25	9.3

surface. This may have potentially affected the shrinkage behavior due to reduced loss of water in the first hours of hydration.

The shrinkage ranking (highest to lowest) obtained from the full-scale environmental sections is as follows: Type II/V, Type III, Type II/V LWA and CSA. This rank is very similar to the one obtained with the free shrinkage beams and suggests that the expected high shrinkage levels for Type III mix were not realistic. The contribution of the LWA could also be different between small laboratory specimens and full-scale slabs.

Analysis provided initial information regarding shrinkage of HESC mixes for rapid BCOA construction. Future complementary reports will expand the analysis related to shrinkage performance of BCOA under accelerated pavement testing.

Strength Performance

Flexural strength is the parameter used by Caltrans for construction specifications. The average results of three specimens for different testing times (opening time, four times opening time and 45 days) were prepared in the laboratory and the summary results are shown in the Table 3.

Average results for specimens prepared in the laboratory for all four mixes fulfilled Caltrans requirement at OT. Specimens were also cast during construction of the test track to check consistency and performance of the concrete being placed. Several constraints were found such as environmental effects (temperature, the wind), hauling and waiting times, and repeatability of the mix coming from the plant. The results of the on-site specimens showed that Type II/V-LWA-10H OT mix didn't achieve the 2.8 MPa (400 psi) threshold and there were some Type III-4H-OT specimens sampled in the second day that also did not achieve the required strength. These results clearly show that better control of production, transportation, and placing of the HESC is needed on the field to achieve consistent and adequate results.

Potential Performance-Related Trade-offs

The differences and trade-offs should be analyzed within certain characteristics of HESC mixtures. Therefore, the following trade-off comparison is based on three main group alternatives that the BCOA mixes provide: portland cement vs. no portland cement (CSA), opening to traffic at four hours vs. opening to traffic at 10 hours and for Internal Curing Concrete (ICC) vs. conventional concrete.

Portland Cement Concrete vs. CSA Concrete. CSA concrete differs from portland cement mix performance in several ways. First, the CSA shrinkage trend shows expansion rather than contraction as occurs with the other portland cement mixes. Second, the CSA mix achieved the Caltrans flexural strength threshold as quickly as 2–3 hours, although the trade-off is lower long-term flexural strength. Production and disposing of the concrete mix is another difference. While portland concrete mixes are produced in the plant and hauled to the site in ready-mix trucks, the CSA concrete mix is typically produced in volumetric mixing trucks and even though the production is completely different the variability is not. Concrete beams were sampled for both CSA and Type III 4-H-OT mixes. Flexural strength tests show that at four hours the average and standard deviation for the CSA mix were 3.35 MPa and 0.16 MPa, while the Type III mix beams had an average and standard deviation of 2.94 MPa and 0.10 MPa.

4-Hours OT vs 10-Hours OT. The type of concrete mix used in a project is determined by opening times. Regardless of that restriction, early strength against long-term strength is an important fact to consider because 4H-OT mixes are able to fulfill specifications in 40% of the time or less than what it takes for the 10-hour mixes. In addition, in the long term, the Type III concrete mix could provide very similar strength performance compared with the two 10-hour mixes. The costs start to play an important role in this relationship because Type II/V cement is less expensive than any of the 4-hour mixes.

ICC vs Conventional Concrete. The third comparison and the respective trade-offs are: using ICC significantly improves the shrinkage and cracking performance. The trade-off in this relationship consists of giving away strength to optimize the hydration process of concrete. The cost of LWA is another consideration.

It is important to remark that these trade-offs cannot be evaluated based only on the mix properties discussed here, and the APT and structural analysis will provide more information to Caltrans and others.

Conclusion

Successful validation and repeatability of mix production was achieved in the laboratory with changes to standard laboratory practices. There is less variability in the data from the laboratory specimens versus the variability of the data from the specimens that were prepared during construction. It is important to control and account for several field factors such as environment, hauling times and construction practices to achieve minimum strength requirements.

The composition of the cement can produce sensitive variations of strength and shrinkage performance. Understanding these differences permits better selection and design for the performance requirements of a given project. The use and doses of admixtures needs to be optimized as the components of the cement and construction conditions vary.

Caltrans' flexural strength approval threshold for HESC mixes agrees with the specification limits of other transportation departments.

In compressive strength, the mix with LWA performs well in the long term, reaching similar values as the 10-hour mix without LWA. However, the short-term and long-term flexural strength performance was less than that of the 10-hour mix without LWA, which might be attributed to aggregate–paste interactions being different when tension stresses are induced in the flexural test. The use of LWA in HESC mixes for quick construction of BCOA needs to be re-evaluated, and more research is needed.

The ratio of flexural strength to compressive strength is similar at OT for all the mixes between laboratory specimens and field specimens. Moreover, the regression chosen fits very well the data for the CSA, Type II/V, and Type II/V LWA mixes. Being able to develop this correlation will ideally allow the use of the compressive test as the performance test to check for Caltrans flexural strength threshold.

The maturity approach does not provide any improvement to the existing time–strength relationship for individual HESC mixes. Further analysis, should be performed to calibrate more accurate maturity models.

The trends of the relatively-free shrinkage beams results were consistent with expectations, although they are not a realistic representation of the constrained full-scale slabs. The ranks in order from the highest to the lowest shrinkage for the environmental full-scale sections were Type II/V, Type III, Type II/V LWA, and CSA. For the relatively-free shrinkage beams the Type III and Type II/V LWA switched places.

A trade-off comparison was performed based on three main group alternatives: Portland cement vs. calcium sulfo-aluminate cement, opening to traffic at four hours vs. opening to traffic at 10 hours, and lastly Internal Curing Concrete (ICC) vs. conventional concrete. Further structural and APT analysis will provide more insight to determine full-scale benefits and trade-offs under traffic.

Acknowledgments

This paper describes research activities that were requested and sponsored by the California Department of Transportation. This sponsorship is gratefully acknowledged as well of the support of Teichert Construction, CTS Cements and Southwest Concrete Pavement Association.

Author Contributions

The authors confirm contribution to the paper as follows: study conception and design: Julio Paniagua; data collection: Julio Paniagua, Fabian Paniagua; analysis and interpretation of results: Julio Paniagua, Fabian Paniagua, Angel Mateos; draft manuscript preparation: Julio Paniagua, John Harvey. All authors reviewed the results and approved the final version of the manuscript.

References

1. Vandenbossche, J. M., and A. J. Fagerness. Performance, Analysis, and Repair of Ultra-thin and Thin Whitetopping at Minnesota Road Research Facility. *Transportation Research Record: Journal of the Transportation Research Board*, 2002. 1809: 191–198.
2. Rasmussen, R. O., and D. K. Rozycki. *NCHRP Synthesis 338: Thin and Ultra-thin Whitetopping*. TRB, National Research Council, Washington, D.C., 2004, pp. 3–9.
3. Zichang, L., N. Dufalla, F. Mu, and J. M. Vandenbossche. Bonded Concrete Overlay of Asphalt Pavements

Mechanistic-Empirical Design Guide (BCOA-ME). *User's Guide, FHWA TFP Study*, Vol. 5, 2013, p. 165.
4. Caltrans. *Standard Specifications in Section 90: Concrete*. State of California, 2015.
5. Caltrans. *Standard Specifications in Section 40: Concrete Pavement*. State of California, 2015.
6. Harrington, D., and G. Fick. *Guide to Concrete Overlays: Sustainable Stubs for Resurfacing and Rehabilitating Existing Pavements*, 3rd ed. National Concrete Pavement Technology Center, Ames, Iowa, 2014.
7. Mindess, S., J. F. Young, and D. Darwin. *Concrete*. Prentice Hall, Upper Saddle River, 2003.
8. Neville, A. M., and J. Brooks. *Concrete Technology. 2010*. Pearson Education Limited, Essex, England, 1987.
9. Whiting, D., M. Nagi, P. A. Okamoto, H. T. Yu, D. G. Peshkin, K. D. Smith, M. I. Darter, J. Clifton, and L. Kaetzel. *Optimization of Highway Concrete Technology*. Report SHRPC-373. Strategic Highway Research Program, Washington, D.C., 1994.
10. Paniagua, J. C., F. Paniagua, M. Angel, and J. T. Harvey. *Development, Characterization, and Performance of Early-High Strength Concrete Mixes for Bonded Concrete Overlays*. University of California Pavement Research Center, 2017.
11. Kosmatka, S. H., B. Kerkhoff, and W. C. Panarese. *Design and Control of Concrete Mixtures*. Portland Cement Association, 2011.
12. Glasser, F., and L. Zhang. High-Performance Cement Matrices Based on Calcium Sulfoaluminate–Belite Compositions. *Cement and Concrete Research*, Vol. 31, No. 12, 2001, pp. 1881–1886.
13. American Concrete Institute Committee. *Guide for the Design and Construction of Concrete Parking Lots*. ACI Report 330R-08. American Concrete Institute, Farmington Hills, Mich., 2008.
14. Raphael, J. M. Tensile Strength of Concrete. *ACI Journal Proceedings*, Vol. 81, No. 2, 1984, pp. 158–165.
15. Zhang, J., J. Harvey, A. Ali, and J. Roesler. *Goal 4 Long Life Pavement Rehabilitation Strategies—Rigid: Laboratory Strength, Shrinkage, and Thermal Expansion of Hydraulic Cement Concrete Mixes*. Partnered Pavement Research Center, University of California, Davis, and Berkeley, 2004.

The Standing Committee on Concrete Pavement Construction and Rehabilitation (AFH50) peer-reviewed this paper (18-05076).

The contents of this paper reflect the views of the authors and do not necessarily reflect the official views or policies of the State of California or the Federal Highway Administration. This paper does not represent any standard or specification.

Sage Publications USA
sagepub.com

ISBN 978-1-5108-7830-3

9 781510 878303

CURRAN ASSOCIATES INC.
proceedings.com